Living in Denial

Living in Denial
Climate Change, Emotions, and Everyday Life

Kari Marie Norgaard

The MIT Press
Cambridge, Massachusetts
London, England

© 2011 Massachusetts Institute of Technology

All rights reserved. No part of this book may be reproduced in any form by any electronic or mechanical means (including photocopying, recording, or information storage and retrieval) without permission in writing from the publisher.

For information about special quantity discounts, please email <special_sales@mitpress.mit.edu>.

This book was set in Sabon by Toppan Best-set Premedia Limited. Printed and bound in the United States of America.

Library of Congress Cataloging-in-Publication Data

Norgaard, Kari Marie.
Living in denial : climate change, emotions, and everyday life / Kari Marie Norgaard.
  p. cm.
Includes bibliographical references and index.
ISBN 978-0-262-01544-8 (hardcover : alk. paper) — ISBN 978-0-262-51585-6 (pbk.)
1. Climatic changes—Psychological aspects. 2. Environmental policy—Citizen participation. I. Title.
BF353.5.C55N67 2011
304.2′5—dc22
                                                                2010036047

10 9 8 7 6 5 4

To the future generations in my family:
To my son, Cody, and to sister Addie, brother Matt, and niece Isabel.
To all future generations.
May your world flourish.

Contents

Acknowledgments   ix
Prologue: An Unusual Winter   xiii

Introduction: The Failure to Act, Denial versus Indifference, Apathy, and Ignorance   1

1 Boundaries and Moral Order: An Introduction to Life in Bygdaby   13

2 "Experiencing" Global Warming: Troubling Events and Public Silence   33

3 "People Want to Protect Themselves a Little Bit": The Why of Denial   63

4 The Cultural Tool Kit, Part One: Cultural Norms of Attention, Emotion, and Conversation   97

5 The Cultural Tool Kit, Part Two: Telling Stories of Mythic Nations   137

6 Climate Change as Background Noise in the United States   177

Conclusion   207
Appendix A: Methods   231
Appendix B: List of People in Bygdaby Interviewed and Quoted   243
Notes   245
References   249
Index   265

Acknowledgments

Despite the individualism we are taught to believe in and follow, few projects are ever completed alone. The people of Bygdaby deserve special thanks for their time, answers, interest, and patience in responding to my questions. If I offer a view of Bygdaby that is critical, it is only because I have posed the hard questions of our times to this particular community, certainly not because people there are more deserving of criticism. Instead, Bygdaby and Norway are important precisely because people there are so sincere in their concerns for the wider world and engaged in so much political activity on its behalf. I am especially grateful to all who shared their thoughts and experiences in formal interviews and casual conversation. Warm thanks to Siren, Knut, Tor, Pål, and the members of the Bygdaby Utferdslag for their friendship and kindness. Thanks as well to the individuals in the United States who shared their perspectives on climate change via interviews, conversations, and examples and especially to the students whose voices enliven chapter 6.

Much support has come from my "more than academic" community. At the University of Oregon, Sandra Morgen, Mia Tuan, John Foster, Jocelyn Hollander, and Paul Slovic provided valuable questions, insight, and reflections from their own standpoints and fields of expertise. The Center for Environment and Development at the University of Oslo generously provided me with institutional support, office space, and a community of colleagues in the early stage of my fieldwork. Barbara, Leontina, Megan, and members of the Social Sciences Feminist Network writing group gave excellent feedback on early chapter drafts, and Dianne Clark and Johanna Stoberock provided substantial assistance in the later stages. I appreciate the anonymous reviewers' encouragement and thoughtful attention and especially the support and time given by my editor Clay Morgan, to Deborah Cantor-Adams, Annie Barva, and others who were instrumental at the MIT Press. Thanks as well to Sarah, Jennifer, and Deborah for lots of great listening and thoughtful support during the writing process.

Funding for my time in the field came from the American Scandinavian Foundation. My colleagues at the University of California, Davis, generously allowed me to continue to think about and work on this project during my National Science Foundation–funded postdoctoral studies. Finally, my truly exceptional colleagues in the Department of Sociology and Environmental Studies at Whitman College provide community, inspiration, and solidarity on a daily basis.

This project would never have happened without the support of a number of less visible people. Back in 1983, my grandmother handed me an advertisement for a study abroad program in Scandinavia and asked if I wanted to attend (I chose Norway because of the mountains). When I arrived, my host family—Sissel, Svein, and Ida Pedersen—welcomed me into their home and community and patiently taught me both the Norwegian language and an appreciation for Norwegian ways of seeing and being that have forever informed my political imagination. Years later in another phase of life, this project took shape. Sam, my partner and husband, went with me to Norway. Sam has engaged with and supported my thinking with questions, observations, and enthusiastic encouragement all the way from my grim beginning in a rainy Oslo "summer" to the joy of skiing from hut to hut across Hardangervidda. Sam shared the struggles with Norwegian formality, dark winters, and unfamiliar language as well as the delights of meeting a new people, culture, and political structures. In the final writing phase, he spent many extra hours caring for and playing with our son, Cody. I am deeply grateful for his engagement with my ideas and his presence in my life. And these days no academic family can thrive on its own. Our family has a set of guardian angels in Gloria and Conrad, who spend time with Cody on a daily basis. I thank them for all their love and energy!

Three of my grandparents died while I was in the process of writing. My last grandparent, my blood link to Norway, passed away just three days before I filed my dissertation with the University of Oregon. My grandparents have been part of my own connection to the past; in their passing, my world is changed. Each of my grandparents has influenced me significantly and thus contributed to this project: getting me to

Norway in the first place, sharing perspectives, and spending time with me as a child. Before his death, my paternal grandfather bought me the computer upon which I wrote. Finally, I thank my parents, Marida and Dick, for years of material and emotional support, for taking me boating down rivers and hiking in deserts and mountains, and for sharing with me their concern for the larger world. Their contribution to this project is more than I can comprehend.

Prologue: An Unusual Winter

During the fall and winter of 2000–2001, unusually warm weather occurred in a rural community in western Norway. November brought severe flooding across the entire region. By early December, it was established that the weather was measurably warmer than usual. The local newspaper reported that October, November, and December were respectively 4.0, 5.0, and 1.5 degrees warmer than the 30-year average. As of January 2001, the winter of 2000 for Norway was recorded as the second warmest in the past 130 years. This fact was highly publicized. Regional and national newspapers carried headlines such as "Warmer, Wetter, and Wilder," "Green Winters—Here to Stay?" and "Year 2000 Is One of the Warmest in History." In the town of Bygdaby[1] (a pseudonym pronounced Big-DAH-bee), where I did my fieldwork, the first snowfall did not come until late January—some two months later than usual.

As a result of these conditions, the local ski area did not open until late December, and only then with the aid of 100 percent artificial snow—a completely unprecedented event with dramatic effects on recreation and measurable economic impacts on the community. The local lake failed to freeze sufficiently to allow for ice fishing. Casual comments about the weather, a long-accepted form of small talk, commonly included references to unusual weather, shaking of heads, and the phrase "climate change." Lene, a businesswoman in her late forties, described the difference in the weather from her childhood:

In my childhood there was lots of snow all the time, it was cold, all the way down to –40 Celsius, so that diesel cars just stopped working, you know? And we had ice on the lake, the kind we had now for a few days. It was like that the entire winter, it was always like that, and we had such a good time. Down at the lake we had music, and there was both a long skating track and in the middle . . . a shorter track. Those were such different times. But since I've grown up, it's been different. We have received little snow. Of course it's wonderful in a way.... You know, you don't have to shovel snow, you don't have to drive on ice, and all that. But the extreme [warm] weather, it didn't come until the 1980s, the end of the 1980s, it seems.[2]

Although the dramatic change in weather may have been most apparent to people older than thirty, teenagers could also perceive that the weather patterns were quite different. Vigdis, a 17-year-old student involved in antiracism work, described the change: "It is, well, milder. There has been less change between the seasons. There is less snow and more, like, halfway winter, and the summers have been colder. I think that it comes from climate change. Because it didn't used to be this way."

In addition to the marked absence of snow, the lake on the edge of town failed to freeze. In late 2000, a woman who was walking on the lake fell through the ice and drowned when it cracked. Ketil, an administrator at a small cultural institute, described the dramatic change in the lake ice over the previous decade:

Like the lake here—until fifteen years ago people came to Bygdaby from eastern Norway, from Hallingdal, and [from] other places by train. They stayed overnight at the hotel in order to use the ice. It was completely black with people out on the ice every single winter. They went out there and fished. It was very good fishing. But you know it hasn't been like that for the last ten years; now it is completely gone. Nobody comes here anymore. It hasn't been safe ice for nearly ten years now. After a day or two, it will rain.

Perhaps the clearest impact of the weather on the community that winter can be measured in money. Because of lack of snow, the opportunity to ski was greatly reduced, and the resort owner had to invest a considerable amount of money and effort to produce a single run made completely of artificial snow.

Communities around the world are experiencing similar stories of unusual weather that seriously impact local economies and survival. Across New Hampshire, a trend toward warmer winters has resulted in fewer and fewer ski areas. The warmer weather has particularly impacted smaller operations, contributing to an industry shift toward larger ski areas (Hamilton, Rohall, Brown, et al. 2003). In Vermont, the month-long season for maple syrup production has decreased by about three days over the past 40 years, leading to measureable decreases in syrup production and syrup producers' worries that climate change has begun to affect the $200 million industry. Communities in polar regions are particularly at risk (Alaska Regional Assessment Group 1999; Arctic Climate Impact Assessment 2005). In October 2004, *Time* magazine ran a story about Shishmaref, a 4,000-year-old Inupiaq Eskimo village on an Alaskan barrier island where the permafrost is thawing and where rising seas threatened to submerge the island. Huge waves had washed away the school playground and $100,000 worth of boats and fishing equipment. Two years later in 2006 the entire community was evacuated (Roosevelt 2004). A recent U.S. General Accountability Office study found that 4 Alaskan villages are in "imminent danger" and that another 20 are seriously threatened by rising sea levels. In fact, the report documents that 184 of 213 Alaskan native villages are presently at risk from serious flooding (US GAO, 2004, 1). Elsewhere in the state, Inuit have difficulty using snowmobiles because the ice is dangerously thin. In the fall of 2004, Inuit people sought a ruling from the Inter-American Commission on Human Rights against the United States for causing global warming and its devastating impacts.

Changing climate has visible impacts farther south as well. The World Health Organization now estimates that worldwide climate change contributes to 160,000 deaths each year due to the increased prevalence of vector-borne diseases, food insecurity, and heat waves (Campbell-Lendrum, Pruss-Ustun, and Corvalan 2003). By 2030, climate change is expected to lead to a 14 percent increase in the number of people exposed to malaria in Africa, and the rate of people at risk from dengue worldwide is expected to double by 2070 (Hales, de Wet, Maindonald, et al. 2002; World Bank 2010). High-income countries are vulnerable as well. The 2003 summer heat wave led to the deaths of more than 70,000 people across Europe (Robine, Cheung, Le Roy, et al. 2008; World Bank 2010). As urban heat islands produce temperatures significantly higher than surrounding areas, city planners are beginning to map patterns. By the middle of this century, New York, Philadelphia, Detroit, Chicago, and Minneapolis are projected to be among the cities in the United States with the most heat-related deaths due to global warming (Carlson 2007).

In the winter of 2000–2001 in Bygdaby, it was not just the weather that was unusual. As a sociologist, I was perplexed by the people's behavior as well. Global climate change is arguably the single most significant environmental issue of our time. Impacts on human society are predicted to be widespread and potentially catastrophic as water shortages, decreased agricultural productivity, extreme weather events, and the spread of diseases take their toll. Potential outcomes for Norway include increased seasonal flooding, decreased winter snows, and the loss of the Gulf Stream that currently maintains moderate winter temperatures, thereby providing both fish and a livable climate in the northern region. In Norway, there has been relatively high public support for the environmental movement as well as public awareness of and belief in the phenomenon of global warming. Yet despite clear social and economic impacts on the community, no social action was taken at the beginning of this century. Whether the warm weather and lack of snow in Bygdaby were actually a result of global warming or not cannot be determined for certain because weather and climate are not equivalent. Among competing explanations for the unusual weather, however, it *was* widely linked to global warming in both the media and in the minds of local residents. National newspaper coverage of weather events contained information on climate change, and small talk about unusual weather frequently referred to the possibility of climate change. In a focus group in late November with young women who attend the local high school, I asked whether the issue of climate change seemed "real" to them or not:

Siri I have heard about the conference [climate meeting at The Hague]. I became a bit afraid when they didn't reach agreement. . . .
Trudi Our minister of environment! In 2008, we will decrease our emissions by 5 percent. (General laughter.)
That will help!
Kari So all of you have followed this a bit. And is it something that you feel is real, really happening, or . . . ?
(Immediately and several speaking at once.)
Mette Now it is incredible, 5 degrees Celsius is, you know, really strange.
(Mmm, ja.)
Siri (interrupting) There should be snow.
Trudi It comes in much closer for us. You notice it. You know, it's getting worse and worse.
Mette We notice it here in the everyday with climate here, in your surroundings.
Trudi Last year there was snow at this time of year. And actually that is the way it should have been for quite some time now.

This conversation occurred on November 28, 2000. The community did not get snow until mid-January.

What perplexed me was that despite the fact that people were clearly aware of global warming as a phenomenon, everyday life in Bygdaby went on as though it did not exist. Mothers listened to news of unusual flooding as they drove their children to school. Families watched evening news coverage of the failing climate talks in The Hague, then just tuned into American sit-coms. Global warming did not appear to be a common topic of either political or private conversation unless I brought up the topic. Aside from small talk about the unusual weather, few people ever seemed to spend much time thinking about global warming.

People could have reacted differently to that strange winter. In Bygdaby, the shortened ski season affected everyone in the community. In the words of one taxi driver, "It makes a difference if we move from five months of winter tourism to only three. It affects all of us, you know, not just those up on the mountain. It affects the hotels, the shops in town, us taxi drivers, we notice it too." Why didn't this awareness translate into social action? Throughout modern history, people have used a variety of strategies to draw attention to problems in their communities, such as staging marches and boycotts and writing letters to newspaper editors and political leaders. What might Bygdabyingar have done differently? Community members could have written letters to the local paper, brought up the issue in one of the many public forums that took place that winter, made attempts to plan for the local effects of climate change, put pressure on local and national leaders to develop long-term climate plans or short-term economic relief, decreased their automobile use, or, at the very least, engaged their neighbors, children, and political leaders in discussions about what climate change might mean for their community in the next ten to twenty years.

Indeed, in other parts of the world that year reactions to climate change *were* different. The severe flooding in England in November 2000 was linked to climate change by at least some of the impacted residents. People from affected communities in England traveled to the climate talks at The Hague to protest government policies. Since that time, several cities in the United States have taken action against the federal government over global warming. And although one cannot tie weather events per se to climate change, the fact that increased hurricane intensity is one clear outcome of climate change has led residents in Mississippi, who are now homeless as a result of Hurricane Katrina, to file a lawsuit against oil companies for their role in climate change. The residents of Bygdaby could have taken similar actions, rallying around the problem of the lack of snow and its economic and cultural impacts. But they did not.

How did people in Bygdaby manage to ignore outwardly such significant risks? Did they manage to ignore it inwardly as well? Why did such a seemingly serious problem inspire so little response?

The rather puzzling behavior of people that winter in Bygdaby is related to larger questions about social and environmental action in Norway, in the United States, and around the world: How are the citizens of wealthy industrialized nations responding to global warming? Why are so few people taking any sort of action? Why do some social and environmental problems result in people's rising up when others do not? And given that many people do know the grim facts, how do they manage to produce an everyday reality in which this urgent social and ecological problem is invisible? Citizens of all the wealthier nations of the world today face these critical questions. Climate change is not unique to Norway, nor are its present and future impacts. Nor, unfortunately, is the failure of response unique to this small community in Norway. Despite the extreme seriousness of this global environmental problem, the pattern of meager public response—in terms of social movement activity, behavioral changes, and public pressure on governments—exists worldwide.

I arrived in Norway in the summer of 2000 on a scholarship from the American-Scandinavian Foundation, with a concern about global warming and an intention to conduct research on how the environmentally progressive Norwegians made sense of it. My husband, Sam, accompanied me with plans to work at the local ski resort. I was fluent in Norwegian. We both looked forward to an interesting year. Norway was not only a place where I had spent significant time growing up, but also a nation I admired for its strong environmental and humanitarian values. Plus, the Norwegians have significant wealth, which can be an asset, at least in making technological changes. Since the time I first lived in Norway as a teenager, I had been fascinated by the extent of progressive environmental policy and awareness there. Now I returned with my comparative sociological lens to ask questions that at the time could not be addressed in my own country, the United States. Indeed, at the time the United States was the only country in the world where, thanks to extensive countercampaigns by the oil industry and the George W. Bush administration, one-quarter of the population still questioned whether global warming was actually occurring (Krosnick 2009).

In the past several years, especially since the widespread viewing of Al Gore's *An Inconvenient Truth* and the events of Hurricane Katrina, the United States has "caught up" with Norway, which is to say that Americans have reached the point that Norwegians were at in 2001: widespread knowledge and concern regarding climate change, but still too little action. More important, having reached the same levels of information and concern that Norwegians exhibited in 2001, many Americans have also begun to experience the same mental landscape inhabited by so many well-educated, environmentally conscious people I met in Bygdaby—a landscape where the possibility of climate change is both deeply disturbing and almost completely submerged, simultaneously unimaginable and common knowledge. In this sense, Norway and the community of Bygdaby serve as a bellwether for the United States and the rest of the world.

People in the United States, facing the same quandary, can no longer claim not to know about global warming. Although some 68 percent of the population list global warming as a serious environmental problem in recent polls, few people spend time writing or thinking about it, much less taking action. A joint study by the American Geophysical Union and Public Agenda in 1998 emphasized the public's feelings of powerlessness and frustration, rather than lack of information, connected to the issue of climate change:

They said they care deeply about global warming, but their concern did not translate into any forward motion. As they thought about the problem, they seemed to run into brick walls, characterized by lack of clear knowledge, seemingly irreversible causes, and a problem with no real solution. As a result they were frustrated and eager for a solution but unsure of which way to go. The symptoms of this frustration are clear. The first is that people literally don't like to think or talk about the subject. Our respondents always seemed to want to move the topic from global warming itself to more familiar topics, such as moral deterioration, where at least they felt on firmer ground. (Immerwahr 1999)

Despite increases in awareness and concern in the United States since the time this observation was made, the comment still holds an eerie familiarity to what I observed in Norway.

This book is about how people experience disturbing information regarding global climate change. It tells a story about what goes on behind the scenes to create the public face of apathy. It is a story that uses the voices of members of one small town to speak to questions from sociology and science communication regarding the relationship between information and social action. It is a story that I hope will help us to understand the complexity of the lived experience of people around the world as we struggle collectively to make sense of this significant problem.

Introduction: The Failure to Act, Denial versus Indifference, Apathy, and Ignorance

Environmental and social scientific communities alike have identified the failure of public response to global warming as a significant quandary. Most existing explanations emphasize lack of information (people don't know enough information; climate science is too complex to follow; or corporate media and climate skeptic campaigns have misled them) or lack of concern (people are just greedy and self-interested or focused on more immediate problems). Such work emphasizes either explicitly or implicitly the notion that information is the limiting factor in public nonresponse to this issue, an approach that is often called the "information deficit model" (see, e.g., Bulkeley 2000). There is the sense that "if people only knew," they would act differently: that is, drive less, "rise up," and put pressure on the government. For example, psychologists Grame Halford and Peter Sheehan write, "With better mental models and more appropriate analogies for global change issues, it is likely that more people, including more opinion leaders, will make the decision to implement some positive coping action of a precautionary nature" (1991, 606). Researchers have lamented the confusion between global warming and the ozone hole (e.g., Bell 1994; Bostrom, Morgan, Fischoff, et al. 1994; Read, Bostrom, Morgan, et al. 1994), investigated the role of media framing (Bell 1994; Ungar 1992; Grundmann 2006, 2007), and described how understanding global warming requires a complex grasp of scientific knowledge in many fields. Also in this vein, John Sterman and Linda Sweeney examine public misperceptions of climate models as a cause for inaction. The authors conclude that "low public support for mitigation policies may arise from misconceptions of climate dynamics rather than [from] high discount rates or uncertainty about the impact of climate change" (2007, 606). Furthermore, they link this misunderstanding to the failure of response by U.S. policymakers. Yet as Daniel Read and his colleagues (1994) pointed out more than a decade ago, only two simple facts are essential to understanding climate change. If significant global warming occurs, it will be the result primarily of an increase in the concentration of carbon dioxide in the earth's atmosphere. And the single most important source of carbon dioxide is the combustion of fossil fuels, most notably coal and oil. How can it be that people don't know these basic facts?

Finally, the information deficit approach cannot explain a paradoxical phenomenon: as evidence for climate change pours in, and as predictions become more and more alarming and scientific consensus increases, interest in the issue in Norway and elsewhere is declining. Biannual national surveys find a significant and steady downward trend in Norwegian interest and concern in the issue, with the percentage of respondents who replied that they were "very much worried" about climate change declining steadily from 40 percent in 1989 to less than 10 percent in 2001 (Hellevik 2002, 13; Barstad and Hellevik 2004).[1] Hellevik's explanation for declining concern is interesting: "A decline from such a high level of anxiety is to be expected. There are limits to how long it is possible for individuals to live with the extremely pessimistic environmental perspectives reflected in the 1989 results. Anxiety reduction mechanisms make people look for brighter aspects of development" (2002, 13). Although the situation is more complicated in the United States, we can see evidence of the same pattern here. For example, Paul Kellstedt, Sammy Zahran, and Arnold Vedlitz have found that increased levels of information about global warming have a negative effect on concern and sense of personal responsibility. In particular, respondents who are better informed about climate change feel less rather than more responsible for it: "in sharp contrast with the knowledge-deficit hypothesis, respondents with higher levels of information about global warming show less concern" (2008, 120). Similarly, Jon Krosnick and his colleagues (2006) observed that people stopped paying attention to global climate change when they realized that there is no easy solution for it. They note that many people instead judge as serious only those problems for which they think action can be taken.

In the United States, there is also the phenomenon of outright climate skepticism, in which 26 percent of the population does not believe there is scientific consensus that climate change is occurring (Krosnick 2009). Is this phenomenon at all linked with the larger majorities of the U.S. public who find global warming alarming, but who fail to take action? If so, how?

Existing studies of how people process information on climate change have focused largely on either the individual level, examining "mental models" and cognitive schemas (e.g., Bostrom, Morgan, Fischoff, et al. 1994), or the national level, carrying out large-scale cross-national surveys (e.g., Dunlap 1998; Saad 2002, 2007 Nisbet and Meyers 2007; Newport 2008; Leiserowitz, Maibach, and Roser-Renouf 2008, 2010). No sociological work to date has taken an open-ended, ethnographic approach to the question of how people experience climate change. Results from the few studies that use interview data do not support the information deficit model. Instead, their results describe a complexity of response, situations of knowing and not knowing, and emotional ambivalence. Perhaps more significant, although information deficit explanations are indispensable, they do not account for the behavior of the large number of people who *do* know about global warming, believe it is happening, and express concern. Outright climate skepticism is flashy and attention grabbing, but survey data make clear that a much larger percentage of the Norwegian (not to mention U.S. and world) population is not skeptical (Hellevik and Høie 1999). If we look closely, these people's inaction becomes an interesting, complex, and, I suggest, important barrier to social change.

Double Realities: Climate Change and Everyday Life

It was not long after my arrival in Bygdaby that I began to sense a paradox. Norwegians are among the most highly educated people in the world. Global warming was frequently mentioned during my time in Bygdaby, and community members seemed to be both informed and concerned about it. Yet at the same time it was an uncomfortable issue. People were aware that climate change could radically alter life within the next decades, yet they did not go about their days wondering what life would be like for their children, whether farming practices would change in Bygdaby, or whether their grandchildren would be able to ski on real snow. They spent their days thinking about more local, manageable topics. Ingrid, a local high school student, described how "you have the knowledge, but you live in a completely different world." Vigdis told me that she was afraid of global warming, but that it didn't enter her everyday life: "I often get afraid, like—it goes very much up and down, then, with how much I think about it. But if I sit myself down and think about it, it could actually happen; I thought about how if this here continues, we could come to have no difference between winter and spring and summer, like—and lots of stuff about the ice that is melting and that there will be flooding, like, and that is depressing, the way I see it."

In the words of one person who held his hands in front of his eyes as he spoke, "People want to protect themselves a bit." These voices are echoed in the United States. One of my female environmental studies students described how "solving global warming seems like such a daunting task, and even I know that it can seem too overwhelming." Another student observed, "Despite my knowledge of the wider climate issues, I am still living the same life."

Community members in Bygdaby described this sense of knowing and not knowing, of having information but not thinking about it in their everyday lives. As one young woman told me, "In the everyday I don't think so much about it, but I know that environmental protection is very important." As a topic that was troubling, it was an issue that many people preferred to avoid. Or as Ingrid put it, "I think that there are lots of people who think, 'I don't have that problem myself; I can't do anything about it anyway.'"

Community members describe climate change as an issue that they have to "sit themselves down and think about," "don't think about in the everyday," "but that in between is discouraging and an emotional weight." People in Bygdaby did know about global warming, but they did not integrate this knowledge into everyday life.

This state of affairs brings to mind the work of historical psychologist Robert J. Lifton. Lifton's (1982) research on Hiroshima survivors describes people in states of shock, unable to respond rationally to the world around them. He calls this condition "psychic numbing." Following his initial studies in Japan, much of Lifton's work has been devoted to describing the effect of nuclear weapons on human psychology, particularly for Americans (see, for example, *Hiroshima in America: Fifty Years of Denial* [1995]). Out of this project, Lifton describes people today as living in an "age of numbing" (1993, 210) due to their awareness of the possibility of extinction (from the presence of both nuclear weapons and the capacity for environmental degradation). In this usage, numbing comes not from a traumatic event, but from a crisis of meaning. Lifton says that all of us who live in the nuclear age experience some degree of psychic numbing. We know that our lives can end at any moment, *yet we live as though we do not know this*. Lifton calls this condition the "absurdity of the double life." We live with "the knowledge on the one hand that we, each of us, could be consumed in a moment together with everyone and everything we have touched or loved, and on the other our tendency to go about business as usual—continue with our routines as though no such threat existed" (1982, 4–5). According to Lifton, the absurdity of the double life profoundly affects our thinking, feeling, identity, sense of empowerment, political imagination, and morality. He writes, "If at any moment nothing might matter, who is to say that nothing matters now?" (1993, 23).

I adapt Lifton's phrase "absurdity of the double life" in coining the term *double reality* to describe the disjuncture I observed that winter in Bygdaby. In one reality was the collectively constructed sense of normal everyday life. In the other reality existed the troubling knowledge of increasing automobile use, polar ice caps melting, and the predictions of future weather scenarios. In the words of Kjersti, a teacher in her thirties at the local agricultural school: "We live in one way, and we think in another. We learn to think in parallel. It's a skill, an art of living." This disconnect between abstract information and everyday life is also reported by Norwegian sociologist Ketil Skogen, who finds that for young people in a rural Norwegian community, "environmental issues in general and global threats like the greenhouse effect in particular, are seen as abstract and irrelevant, and are generally not something young people think about" (1993, 232).

It can be easy to take such statements at face value, and most people do. But through his work on the sociology of cognition, Eviatar Zerubavel reminds us that society teaches us what to pay attention to and what to ignore. We learn "cognitive traditions" through a process of socialization. Deciding whether to pay attention to a given idea or event in a given moment or not is a learned process that Zerubavel calls "optical socialization": "Separating the relevant from the irrelevant is for the most part a social act performed by members of particular 'optical' communities who have been specifically socialized to disattend certain things as part of the process of adopting the distinctive 'outlook' of their community. In other words, we learn what to ignore, and only then does its irrelevance strike us as natural or 'logical'" (1997, 47).

Zerubavel tells us that rather than taking thinking as matter of fact, we need to realize that notions of what to pay attention to and what to ignore are socially constructed. We learn what to see and think about from the people around us. Zerubavel's work tells us that whether people notice information about climate change is related to socially shaped systems of perception and attention, whether they remember what they hear is a function of social systems of memory, whether it is considered morally offensive or not is a function of whether it is inside or outside socially defined limits of concern; and the relevance of climate change to daily life is a function of socially shaped systems of cognitive organization (see Zerubavel 1997). "Cognitive traditions" or collective patterns of thinking differ from one "thought community" to another. How we think is part of culture and marks our participation in community. Cognitive traditions and thought communities thus shape how and whether groups of people think about climate change and whether they perceive the topic as relevant for everyday life. From the inside, boundaries of thought appear "natural," and "commonsense" decisions about what to pay attention to or ignore appear strange only when we are outside a given cognitive tradition. Zerubavel (2002, 2006) calls this social shaping of our awareness, memories, and thought patterns the "social organization of denial." Most research to date has examined denial on the level of individual psychology. Yet what individuals choose to pay attention to or to ignore must be understood within the context of both social norms shaping interpersonal interaction and the broader political economic context. Thus, Zerubavel argues, and I agree, that we need both psychology and sociology to study "the mental processes of attending and ignoring" (1997, 11). From the former perspective of individual psychology, people block information on their own as individuals, but from the latter perspective denial occurs through a process of social interaction.

Zerubavel also calls our attention to the normative aspect of how we direct our awareness. Indeed, in every community there are social rules for focusing attention, including rules of etiquette that involve tact-related ethical obligations to "look the other way" and ignore things we most likely would have noticed about others around us. "Not only does our social environment provide us with a general idea of what we can disattend, it very often also tells us what we should repress from our consciousness and ignore. In other words, there is an important (though relatively unexplored) normative dimension to relevance and irrelevance. Indeed, probably the main reason that our own focusing patterns seem so natural or 'logical' to us is that they are usually normatively binding" (Zerubavel 1997, 50).

But why would thought communities be normative? And if they are, then how are the boundaries enforced? Questions about whether people pay attention to climate change can suddenly start to look much like theoretical questions about the nature of power. In the midst of whether climate change is defined as near or far, relevant or irrelevant, we find entree into the heart of concepts such as hegemony and ideology and into the role of culture in the reproduction of power.

Ann Swidler's Cultural Tool Kit and the Production of Culture

One of Antonio Gramsci's (1971) key contributions to social theory is his emphasis on how social control is enacted through the acceptance of ideas that prevent social change and on the important role of culture in legitimating those ideas. If we entertain for the moment the notion that power may be located in the realm of culture, then we must next ask both *how* and *why* particular systems of memory or cognition concerning climate change are enforced. How exactly does power operate through culture? Up until the mid-1980s, many social scientists understood culture to shape human activity in a fairly static manner, through providing values that direct actions.

Then in 1986 Ann Swidler's work described an alternative framework for a causal role of culture in social action. In a groundbreaking essay, Swidler describes how "culture influences action not by providing the ultimate values toward which action is oriented, but by shaping a repertoire or 'tool kit'" (1986, 273). In her view, culture shapes social action not by providing guiding values, but by providing cultural components or "chunks of culture" (283) that can be used as tools by individuals to construct "strategies of action" (273). Such a "tool kit" may contain "symbols, stories, rituals and world-views which people may use in varying configurations to solve different kinds of problems" (273). For example, "Publicly available meanings facilitate certain patterns of action, making them readily available, while discouraging others" (283). For Swidler, "This revised imagery—culture as a 'tool kit' for constructing 'strategies of action,' rather than a switchman directing an engine propelled by interests—turns our attention toward different causal issues than do traditional perspectives in the sociology of culture" (271). I build on Swidler's tool kit concept in chapters 4 and 5.

"We Don't Really Want to Know": Climate Change and Disturbing Emotions

A second question about power and culture concerns *why* individuals choose to enact cultural systems of optical socialization. In the fall of 2000 in Bygdaby, using ethnographic fieldwork, interviews, and media analysis, I began to notice that although global warming was an issue that people knew and cared about, they didn't seem to *want* to know about it. Furthermore, not wanting to know about climate change appeared to be related to the host of powerful emotions the topic engendered. The people I interviewed described fears about the severity of climate change, of not knowing what to do, that their way of life was in question, and that the government would not adequately handle the problem. They described feelings of guilt for their own actions and the difficulty of discussing the issue of climate change with their children. These emotions were significant. What role might they play in the equation of the double reality?

In her landmark book *The Managed Heart: Commercialization of Human Feeling* (1983), sociologist Arlie Hochschild vividly exposes the relationship of seemingly private and personal emotions to social structure and political economy. She writes about the signal function of emotions and their ability to provide information about our interpretations of the world. Quite provocatively and in contrast to the emotion/reason dualism, she writes that emotion "can tell us about a way of seeing" and that "emotion is unique among the senses because it is related to cognition" (1983, 220). She goes on to explain: "A black person may see the deprivation of the ghetto more accurately, more 'rationally' through indignation and anger than through obedience or resigned 'realism.' He will focus clearly on the policeman's bloodied club, the landlord's Cadillac, the look of disapproval on the employment agencies' white face. Outside of anger these images become like boulders on a mountainside, miniscule parts of the landscape" (30n.). Hochschild writes that "a person totally without emotion has no warning system, no guidelines to the self-relevance of a sight, a memory or a fantasy. Like one who cannot touch fire, the emotionless person suffers a sense of arbitrariness, which from the point of view of his or her self-interest is irrational. In fact, emotion is a potential avenue to 'the reasonable view'" (30). Sociologists of emotion also emphasize the role emotions play in the sociological imagination: "Emotions provide the 'missing link' between 'personal troubles' and broader 'public issues' of social structure, itself the defining hallmark of the sociological imagination" (Williams and Bendelow 1997, xvii). Thus, Hochschild notes, "When we do not feel emotion, or disclaim emotion, we lose touch with how we link inner to outer reality" (1983, 223).

Scholars such as James Jasper have expanded our understanding of the important role emotion plays in public life and social movements. Outrage, for example, can lead to protest and thus to social change.

Emotions are tied to the moral values that are part of a social movement framing process, shape social movement goals, provide motivation for potential participants to enter movements, and form the basis of solidarity among movement participants (see Jasper 1998; Goodwin, Jasper, and Polletta 2001). The emotions people described in Bygdaby that fall and winter of 2000–2001 were not trivial. If emotion and cognition are linked, it would seem important not to skip over them. Yet few scholars have paid attention to emotions and climate change or to the role of emotion in preventing action. Were these emotions part of the matrix of ignoring? If so, then how?

In contrast to research that has emphasized individuals' faulty decision-making powers; the use of inappropriate schemas, heuristics, or mental models; or the inadequate transfer of information from scientists to society (as in risk and "science communication" models), I aim to shift our view to the emotional and psychological experiences of noticing or thinking about climate information, the normative aspects of thinking and feeling, and the active production of cultures of emotion and talk regarding climate change. In so doing, I am building a model of socially organized denial. In sharp contrast to psychological approaches to denial, the notion of socially organized denial emphasizes that ignoring *occurs in response to social circumstances* and *is carried out through a process of social interaction*. I am grateful here for the work of Eviatar Zerubavel (2002, 2006), who, as previously noted, coined the phrase "social organization of denial" and describes numerous important components of it. Building from the ground up with ethnographic and interview data as this project does, emotions and culture also emerged as prominent factors in the production of denial regarding global warming. Thus, adding to Zerubavel's framework other theory from sociology of culture and emotions, I describe socially organized denial as the process by which individuals collectively distance themselves from information because of norms of emotion, conversation, and attention and by which they use an existing cultural repertoire of strategies in the process.

Denial is also related to political and economic circumstances, what Marxist scholars call "political economy." Norway is one of the nations of the world that has benefitted most from oil production. It is important to understand not only that Norway is one of the world's richest countries, but that oil and gas have played a significant role in generating that wealth. As of 2009, Norway was the world's fifth-largest oil exporter and the second-largest exporter of natural gas (United States Energy Information Administration 2009). As a result of the total volume produced, direct government ownership, and the taxation scheme, more than 34 percent of national revenues came from the petroleum industry in 2008. At the close of 2009, Norway's State Petroleum Fund was worth 2.6 trillion Kroner, or more than 457 billion U.S. dollars.

High levels of wealth, education, idealism, and environmental values, together with a petroleum-based economy, make the contradiction between knowledge and action particularly visible in Norway. The country has moved from a position of environmental leadership (setting a goal of national stabilization of carbon dioxide emissions) to one of political and economic conservatism. In the 15 years prior to my time in Bygdaby, it dropped the goal of capping national carbon dioxide emissions, expanded petroleum development, participated in the Umbrella Group,[2] led the way in developing strategy for trading rather than reducing emissions, and justified increased carbon dioxide through shifting the dialog to an international rather than national context. Oil and gas production increased threefold in the ten years prior to my time in the field. In 2008, the oil and gas industry accounted for 26.6 percent of the national carbon dioxide emissions (Klima og Forurensnings Direktoratat 2009). Norwegian scholars Eivind Hovden and Gard Lindseth write that "Norway, an already wealthy and highly developed country, built a very significant fortune in the 1990s from the very activity that has made stabilisation of CO_2 emissions next to impossible" (2002, 163). Emotions of guilt and powerlessness articulated by the individuals I met must certainly be understood in the context of these political economic conditions.

The term *denial* is sometimes used to describe the phenomenon of outright rejection of the notion that certain information is true—which, in this case, is the reaction of global warming skeptics mentioned earlier. But by now it should be clear that this use of the term is very different and more literal than my use in this book. In his recent work on denial, British sociologist Stanley Cohen (2001) describes three varieties of denial: literal, interpretive, and implicatory. His framework is useful in explaining this book's particular focus. Literal denial is "the assertion that something did not happen or is not true" (the global warming skeptics). In interpretive denial, the facts themselves are not denied but are instead given a different interpretation. Euphemisms, technical jargon, and word changing are used to dispute the *meaning* of events—for example, military officials speak of "collateral damage" rather than the killing of citizens. It is Cohen's third category, implicatory denial, that is at the center of this book. In the case of implicatory denial, what is minimized is not information, but "the psychological, political or moral implications that conventionally follow" (2001, 8). What I observed in Bygdaby—indeed, what we all can observe in the public silence on climate change in United States and around the world—is not in most cases a rejection of information per se, but the failure to integrate this knowledge into everyday life or to transform it into social action. As Cohen puts it, "The facts of children starving to death in Somalia, mass rape of women in Bosnia, a massacre in East Timor, homeless people in our streets are recognized, but are not seen as psychologically disturbing or as carrying a moral imperative to act. . . . Unlike literal or interpretive denial, knowledge itself is not at issue, but doing the 'right' thing with the knowledge" (2001, 9).

My work here draws on theory from diverse academic traditions. Research in the field of environmental sociology and risk perception points to the social and political significance of climate change as well as to the paltry public response it has received. The work of historical psychologist Robert J. Lifton contributes the concepts of psychic numbing and the double life. The field of sociology of emotions describes links between thinking and feeling as well as the process of emotion management, and it asks questions about the role of emotions in social movement participation and nonparticipation (Hochschild 1983; Scheff 1988, 1997; Bendelow and Williams 1998; Goodwin, Jasper, and Polletta 2001). The sociology of culture examines the role of conversation in the development of a sociological imagination and thereby of political power (Eliasoph 1998); how culture can provide resources for constructing strategies of action (Swidler 1986); and how attention, memory, and focus are socially organized (Zerubavel 1997, 2002). Social psychology adds insight on identity and social cognition, describing the power of cognitive schemas and the desire to maintain a sense of self-efficacy and to view self in a positive light. The work of Gramsci and other Marxist theorists on ideology and hegemony highlights links between material relations and dominant discourse, how the ideas of elite political and economic actors come to be seen as common sense to the general public, and how control in modern societies is maintained though consent to "ruling ideas" rather than through direct imposition of force. Using Swidler's concept of culture as a "tool kit" of available resources, I describe how members of Bygdaby had available what I call "tools of order" and the "tools of innocence" to create distance from responsibility, to assert rightness or goodness of actions, to maintain order and security, and to construct a sense of innocence in the face of the disturbing emotions associated with climate change (see chapters 3, 4, and 5).

Weaving these pieces together, I follow an arc of power that moves from the microlevel of emotions to the mesolevel of culture to the macrolevel of political economy and back again. According to my data both from Norway and from the United States, thinking about global warming is difficult for community members because it raises troubling feelings, feelings that go against a series of cultural norms. And these norms are in turn embedded in the particular social context and economic circumstances in which people live. For example, only by analyzing cognition within the context of political economy can we explain Hanno Sandvik's (2008) provocative finding that a nation's willingness to contribute to reductions in greenhouse gas emissions is inversely related to both emissions and national wealth. Thus, in contrast to psychological and survey research that studies human perceptions of climate change on an individual level, I locate these emotional and psychological experiences in both *cultural* and *political-economic* contexts. As a result of this emphasis on cultural, economic, and social contexts, my approach shifts from an "information deficit" model, in which the public fails to respond because of a lack of information, to a "social organization of denial" model in which the public on a collective level actively resists available information.

This book is about how people experience the reality of global climate change in their everyday lives. Most of it is based on an ethnographic account from one community in Norway (see appendix A). But climate denial is not unique to this one community. As an American, I have observed similar dynamics in my own country. In order to show the broader salience of the voices from this one community, including their similarities and differences with voices in the United States, I draw in particular on a smaller set of U.S. interviews, national surveys on climate change, and observations and data collected by other U.S. scholars. Chapter 6 gives a view into how many of the themes developed from Bygdaby play out in the United States. This book is not an exhaustive attempt. I use the comments that people in one community in Norway made and my own observations of them during a recent very dry and warm winter in order to make visible the private emotions and cultural constructions that tell a larger story behind worldwide public paralysis in the face of predictions from climate scientists.

1

Boundaries and Moral Order: An Introduction to Life in Bygdaby

> We think not only as individuals and as human beings but also as members of particular communities with certain distinctive cognitive traditions that affect the way we process the world in our minds.
> —Eviatar Zerubavel, *Social Mindscapes*

Culture and everyday life organize people's experience of our world (including larger global events such as climate change) in different ways. Some features of culture and everyday life—such as the belief in science, high newspaper readership, and the prevalence of political talk—facilitate thinking about climate change. To this extent, it makes sense to speak of the social organization of *awareness* of global warming. From a human ecological standpoint, however, in which we understand ecological and human systems to be threatened by anthropogenic climate change, it is our current *human failure to respond* to global warming to date that holds most significance. Humans have evolved in ecological context. As environmental conditions change, species respond. It is thus critical from a human ecological standpoint to understand why people, those in affluent countries especially, are *not* thinking or taking action about climate change. In this chapter, I identify how features of culture and everyday life make it harder for people to think about climate change and easier for them to ignore it—what Eviatar Zerubavel (2002) calls the "social organization of denial."

In many ways, the people of Bygdaby are similar to citizens in other Western nations: they eat pizza, use cell phones, watch Hollywood movies on TV, believe in democracy, drink cappuccinos, and drive cars. In other ways, they are very different. Norwegians are among the most educated and environmentally minded people in the world. They are also among the wealthiest, yet their wealth is distributed more equally than in most countries. If we want to see how real people in a real community make sense of global warming, we need to get to know them a little better.

What do people in this community spend their time thinking about? Do they get involved in political issues or tend to sit by the sidelines? How do they understand themselves? How do they create a sense of community? Do they think about events in the future, such as climate scenarios? How do they relate to people in other places? What are community residents proud of? Afraid of? The "cognitive traditions" that Zerubavel describes emerge from shared everyday life practices over time, through the collective lived experiences, habits, and meanings we come to understand as culture. What are Bygdaby's cognitive traditions, and how do they inform individual and collective responses to the issue of global warming?

"This Is Bygdaby!"

On our third night in Bygdaby in 1999, my husband, Sam, and I attended a party focused around a rather unusual seasonal food-preparation activity: the traditional preparation of sheep's heads. Six of us were gathered in the cold around a 55-gallon drum upon which sat half-a-dozen sheep heads. Arne used a blow torch to burn the hair off the heads while the rest of us watched. I was playing the roles of sociologist and curious American, doing my best to mask the aversion I felt in reaction to the sights and smells before me: "How many will you make?" "Why do you do it in the fall?" "How long have people prepared sheep heads this way?" "What else is unique about your town?" When Torstein picked up one of the sheep's heads, he pronounced emphatically, "This is Bygdaby!" It was clear we were in for an interesting time.
—Field notes, September 5, 2000

Bygdaby is a town of about 10,000 to 14,000 people, depending on how far from the town center the boundaries are drawn. The town lies in the valley of the Bygda River (another pseudonym, of course). On all sides, mountains slope upward to the surrounding high country. A small and densely packed downtown area of two main streets is clustered around a church built in 1270. When Bygdaby was bombed by the Germans during World War II, most of the downtown was destroyed and so had to be rebuilt. Two- and three-story buildings feature shops at street level, with offices and residences in the upper levels. Older buildings that survived the bombing are interspersed with newer structures. The sidewalks are cobblestoned in places. The downtown features nine or ten "corner grocery stores" (each with nearly identical contents), two banks, a post office, two larger hotels and several smaller ones, a movie theater, the city hall, several schools, and seven or eight cafés, each with its own history, atmosphere, and clientele. On the north side of town, tiny streets wind through neighborhoods, each with its own name and history, until the houses back out against the steep mountain flank. The downtown area is bounded on the west by the lake and on the east by the river and bridge. Packed in between mountains, lake, river, and farms, the town center of Bygdaby is compact. Most residents have an easy walk to downtown. In the words of one woman, "We are Norwegians, we use our legs." Indeed, pedestrian traffic is common as people walk home from one of the many local stores carrying their grocery bags. Buses and trains connect the local region with surrounding communities several times per day.

Just outside of town, on a strip of land beside the river, stand the buildings where occupying Nazi soldiers lived during World War II. Beyond the river is a small section of houses, and beyond that begin the farms: red, white, and yellow buildings set amid small fields. The farms extend upward on the gently sloping lower side of the mountain until the trees and steeper slopes make farming too difficult. Then the mountains extend for hundreds of kilometers in all directions.

Bygdaby is primarily an agricultural area of single-family farms, but it also has secondary industries connected to commerce, communications, education, and tourism. Farms stay in families for generations, traditionally being passed down from father to eldest son. The demographics of Bygdaby are affected by the fact that it is the regional center of the large but sparsely populated surrounding rural district. Many young people (ages 15–18) live in Bygdaby during the school week and return home to their smaller villages in the mountains and along the fjords on weekends. Older people from the surrounding district who cannot continue to live alone on their farms come to Bygdaby for medical care and to retire.

Features of daily life in Bygdaby, from the lack of poverty to employment patterns and lines of political allegiances, are a reflection of the fact that Norway is a social democracy with one of the most comprehensive welfare states in the world. The period from 1945 to the mid-1970s was the "golden age" of the Norwegian welfare state. Social democratic egalitarian ideals were fundamental to Norwegian welfare policy regarding health, education, gender, and equality between the rural and urban regions. In this same period, major improvements were made in Norwegians' daily standard of living: "There was full employment,

Figure 1.1
Farms on the outskirts of the community of Bygdaby.

working hours were reduced, and the electrification of the country was completed. Housing standards rose, and the number of private cars per capita trebled between 1960 and 1974" (Mjøset 1993, 116). Throughout this time, the Labor Party essentially ran the country. Aside from a single three-week period, Labor held the government from 1945 to 1965. Norwegians are proud of the level of equality among citizens that these policies have achieved. As a result, most of the houses in Bygdaby are attractive and relatively similar in size, quality, and status. Three- and four-bedroom homes are most common, well made of wood in roughly the same style. Some people—mostly students and retired people—live in apartments. There are also just a few much nicer and newer houses high up on the hillsides near the ski area and farther outside of town.

What goes on in the minds of people in this quaint setting? What do residents care about? One suggestion for why people fail to respond to global warming is that they are not engaged politically in *any* issues. Such an answer may hold more ground in certain regions of the United States than in this community in Norway. People in Bygdaby are remarkably politically active. Events from marches to local party meetings are announced in the classifieds section of the local paper. During the

11-month period of my stay, community members in this smallish town marched in the streets to mark the anniversary of Crystal Night, picketed over the closing of a local slaughterhouse, and marched in a parade for May 1, Labor Day. Issues on display in the May 1 parade included opposition to the European Union, genetically modified foods, and racism. There were also community forums against the development of a new shopping center and discussions of city planning practices.

High levels of political activity in Bygdaby follow the larger national pattern, from voting to participation in local political parties, involvement with grassroots organization, and street protests. National voter turnout in Storting (National Parliament) elections was 78 percent in the 1997 election and has been around the 80 percent level since the 1970s. Voter turnout in the last local election was 60 percent. Levels of political alienation are lower than in most other Western democratic nations. A comparative study of Norway, Sweden, and the United States measuring feelings of alienation from government over the period from 1968 to 1996 found that whereas trust in the government declined dramatically in Sweden and the United States, it dropped only slightly in Norway. The authors note that "somehow Norway escaped from the more enduring deterioration in support which has affected both Swedish and American politics" (Miller and Listhaug 1998, 164). Norway has 21 registered political parties, 8 of which are active in Bygdaby: the Labor, Center, Socialist Left, Progressive, Red-Electoral Alliance, Christian Democratic, Liberal, and Conservative parties.

Bygdaby has a strong sense of local pride, or *lokalepatriotisme*, that is expressed in their fight against "centralization" (a common theme throughout rural Norway) and in local language politics. There is a palpable tension between a focus on daily life and tradition within this community and the pressures exerted by the wider world (perceived as both Norway as a whole and other nations). This tension plays out in terms of the orientation many Norwegians have to the European Union, some suspicion of strangers, and a fierce retention of the local dialect.

Despite its small size, Bygdaby supports an impressive number of sports organizations. Even more impressive to an American is the percentage of these organizations that focus around the different types of skiing. Bygdaby has a reputation for producing good skiers—during the 1994 winter Olympics in Lillehamar, six athletes on the Norwegian National Team were from Bygdaby. Per, a taxi driver who took me up to the ski area one morning, explained that Bygdaby has lots of good skiers apparently because there are so many farmers. Farmers are

Figure 1.2
Opposition to the European Union. This banner says "Yes to Democracy, No to the EU." From the May 1 parade in 2001.

considered to make good skiers because they grow up using skis in winter from a young age. Indeed, skiing is a prevalent and traditional activity, and the red barns that dot the hillsides seem to contain dozens of pairs of skis. The barn on the farm where my husband and I lived contained half-a-dozen hand-carved wooden skis from the 1800s and dozens of newer skis in all varieties—some designed for ski jumping, others for cross-country skiing, racing, and telemark skiing.

Bygdaby has a small downhill ski area and lighted cross-country tracks and receives some income from ski tourism in winter, although, as noted in the preface, low snowfall for the past ten years has caused income from ski tourism to decrease. The lack of recent winter snow is an issue of concern for both the culture and the local economy, as Ketil, the town's cultural administrator and a longtime resident, explained:

Ketil And things are changing now. So therefore more emphasis is placed on tourism, without getting the degree of employees that one would have wanted.
Kari Tourism hasn't provided that?
Ketil No, it hasn't been as good as it could be, and that is because of the climate that we have had here for the last ten years.

Boundaries and Moral Order 19

Figure 1.3
Old skis in a Bygdaby barn.

Kari What do you mean?
Ketil We haven't had snowy winters so that we could use the ski slopes and gain from ski tourism the way that we wanted to.

The warmer winters affected the economy in another way as well. In the past, there was winter tourism from ice fishing, but that no longer happened. Outdoor activities have been described as an important feature of cultural life for Norwegians. Norwegian philosopher Gunnar Skirbekk says that Norway is a "state that to a great degree builds its national identity on nature. . . . We ourselves are small and vulnerable, and we must understand that we do not stand outside of nature as all-powerful engineers, but that we belong to nature, as a part of the whole" (quoted in Reed and Rothenberg 1993, 6). Christen Jonassen's study *Value Systems and Personality in a Western Civilization: Norwegians in Europe and America* opens by emphasizing the importance of the Norwegian landscape in shaping Norwegian culture:

The Norwegian natural habitat has had profound meaning for Norwegians and has influenced their society, culture, and personality, as demonstrated by the works of its poets, novelists, and dramatists, whose quest for the wellspring of Norwegian psyche and spiritual and emotional life has led them constantly to recognize the natural features of their country. This reaction is understandable to anyone who has experienced that nature firsthand; it is impressive and defies neglect, even as Norwegians put ever greater distances between themselves and nature by the rapid development of their technology and cities. (1983, 3)

Especially attuned to their natural environment, Norwegians are in a special position to notice climate change and to have their national identity affected by it. What happens to such an identity when the environment it derives from begins to change? What happens to a ski culture when there isn't snow?

A strong connection to nature was evident among longtime residents of Bygdaby, or *Bygdabyingar*, as the people of Bygdaby would be known in Norwegian. During interviews, I often asked people what they liked most about Bygdaby. The most frequent answer was the proximity to the mountains. Marit, a woman in her midthirties, gave a typical answer:

Marit Well, I don't know what to say, but it is about the right size—in a way. In a way it is a city, and also it's very rural too. And we have many beautiful mountains here. And that's something that means a lot to us. We like the mountains a lot.
Kari And then you go hiking?

Marit Yes. We go hiking, yes. We have these sheep, you know. We have them up in the mountains in the summer, and then it's just natural that you need to go up and check on them, you know. And so you just love the mountains. So that's something I would totally miss if we should move to the East or something like that. I think that's the thing that would be the worst.

In addition to the value placed on the natural world, mountains, and snow, there is a strong emphasis on maintaining the traditional way of life in Bygdaby, with its close connection to nature. Tradition is very visible in the community, and traditional practices organize time, space, and sense of community. As Kjell Magne, a politically active man in his late forties, put it, "We are concerned with culture in a way, it is completely true. Especially here in Bygdaby, more than [in] many other places that I know. Here in Bygdaby we are really good at taking care of culture and building on old traditions. We really want things here in Bygdaby to be the way they are."

With its mountains and skiing, its hiking and farming culture, and its rich traditions and sense of community, Bygdaby seems very idyllic, and in many ways it is. But it is also a place where people every day face changes and challenges to the traditional way of life they work so hard to maintain. One of the most tangible changes the community is experiencing is the warmer climate, which materially affects the local culture and economy. But how and how much do community members register and react to climate change?

Vulnerability and Security

Information from climate scientists about global warming is highly disturbing. It challenges the basics of our social organization and raises questions about future quality of life and human survival. Our reactions to such information touch on tensions between vulnerability and security in daily life. In Bygdaby, there was a particularly pronounced tension between the known, ordered, safe universe of the town and the larger world around it. There was a desire to keep change out, including climate change, but nonlocal problems were sometimes hard to ignore. Residents were aware, for example, of the economic and political struggles of their close neighbors in Russia and of the poverty and political oppression in other parts of the world. Such information posed a challenge. Liv, a woman active in international human rights work, described the contrast between the safety of her life in Bygdaby and her awareness of the turmoil just outside Norway's borders: "We sit here so safe and good; even if we have unrest around us here, too, you know it's like a nursery school compared with some places. Just think of our neighboring nation, Russia, how they have it!"

Reactions to threats from the outside world can be strong. During the winter of our stay, there was a great scare over hoof-and-mouth disease. To the south in Belgium, thousands of cows and sheep were slaughtered in an attempt to control the spread of the disease. Farmers in Bygdaby were worried. Buckets of chlorine were placed at the door of barns, and only minimal personnel went in and out, carefully cleaning their feet on the way. Torstein, the farmer on whose property we lived, asked us concernedly whether we planned to have any visitors from out of the country. The English schoolchildren who normally came to Bygdaby for their ski holiday were asked to stay home.

Not so long ago another outside threat hit Norway: radiation from Chernobyl. Like global warming, radiation from Chernobyl disregarded national borders. Although most radiation ultimately fell farther north, the entire country of Norway was affected, and at the time of the event nobody knew what would happen. When the conversation in an interview turned to environmental issues, several community members mentioned this event. When I asked Kjell Magne what his concerns were for Bygdaby and Norway in the future, he mentioned the possibility of "another Chernobyl."

The thing I am most afraid of in the entire world, that I can easily say—that is that the Soviet, Russia, becomes so poor that they cannot take care of all their nuclear power plants. I think actually that it is quite possible that it could happen at any time that we have another Chernobyl accident that affects all of us. So that is what I am afraid for today. They don't have a good enough economy to maintain things and do what they need to take care of everything.

Another community member, Sara, described the Chernobyl event to me in vivid detail:

Suddenly it didn't help if you had given any money or not, there was just a cloud coming. Then we were told, I still remember, that [there was] a cloud of radiation that nobody knew where it would end up. It had dangerous things in the air and all, right? And the consequences we have had for many years with cancer, as we have seen. And we have seen that there has been quite an increase. . . . And then I realized for the first time how serious things were. You can't stop it with pretty words, you can't stop it with money, nothing. Here you have to come into dialogue, really, with those who are in charge. So nobody knew where most of it would end up. Not at that time. They have been able to make a map later, you know. I remember myself that I sat at a party, and then we got the message, and people just felt this angst. What in the world is happening now? How serious is it? We had heard about radiation in the meat and so forth. Lots of animals and sheep had to be killed, so that we have been affected by that. The consequences have been huge. And we have seen those who lived in the area, how they look. Genetic alterations and all these things that make us suspicious. . . . There is a lot, I think, somewhere up there in Trøndelag, [and] I wonder if there haven't been serious consequences with lots of cancer and the like.

The tension between vulnerability and security, in Bygdaby as elsewhere, is grounded in memory. The memory of Chernobyl makes many residents nervous, and older *Bygdabyingar* remember well the time when their community was not the safe, secure place it appears to be today, a time when the downtown was bombed and Nazi soldiers occupied the area. These tensions between vulnerability and security are spaces community members must navigate to create the sense of a safe world.

Forming Community

Especially during times of change and crisis, people create a sense of community in part by what they agree to pay attention to and ignore. Being part of Bygdaby is very much about shared cognitive traditions, shared notions of what to think and talk about, what to feel—in short, all that is involved in the collective maintenance of reality. As we will see, residents' complicated reactions to local effects of climate change are directly linked to their efforts to maintain a traditional lifestyle. The community's very close-knit nature makes understanding the social interactions there especially important.

The *Bygdabyingar's* social networks have the mark of tradition. Friendships are based on long-time relationships that involve work and leisure, usually between neighboring farms. The pattern of these relationships among families on neighboring farms can continue for generations. In the past, the practice of *dugnad*, or volunteering one's time to the community, also followed these spatial patterns. *Dugnad* was organized around neighborhood associations, or *grendelag*. These neighborhood associations worked together on tasks from building construction to road repair and handwork projects. Neighborhoods are significant organizational units for friendships and volunteer networks, even in a place as small as Bygdaby. Members of each neighborhood, or *grend*, form groups to work on specific tasks. As Liv described, "People are really active with things. And we live in these little groupings, you can say, here and there. At Fossen, they have the Sanitetsgroup and Skulestad.

Very much *grend* organizations. They gather together quite a bit. Then they do volunteer projects and have a party together. They celebrate St. Han's [midsummer night] together, and it's like this with companionship; it is very important."

This focus on the local affirms the importance of place. My husband and I participated in several of these types of activities. In the spring, the barn on the farm where we lived was condemned and needed to come down. One morning in May, eight to ten of the local (male) farmers came by to help take down the barn. Six tractors were put to work, and the barn came down in less than an hour. People worked together to separate out wood to be reused, gathered smaller pieces into piles, and used chainsaws to cut up the wood for burning. At the end of the day, when I came back from a trip to town, I found the six farmers gathered in the basement drinking home-brewed beer and telling stories.

Community is also expressed through maintenance of local traditions around food and other activities. Certain foods are eaten on certain days of the week and in certain seasons. Many people routinely prepare and eat traditional foods such as *hakasteik* (a porridge made from pork, mutton, and barley) and *raspekaker* (a kind of potato pancake); wear national costumes, or *bunads*, on significant days; and keep to their traditional dialect. Bygdaby has its own special *bunad* distinct from the costumes of nearby valleys. These costumes are worn by many people on special occasions: weddings, church confirmations, and Constitution Day, May 17.

Finally, the cultural norms that serve to bring people together are enforced in Bygdaby through practices of talk. There is a keen suspicion of different behaviors, as described more fully later. Those who do things differently are the subject of much (generally negative) talk and attention. All of this talk preserves a sense of collective identity.

The social boundary between those who are and are not from Bygdaby is anything but subtle. Farmers and old-time Bygdabyingar are perceived as being a tight group, doing things their own way, not listening to or holding much regard for "outsiders." The people of Bygdaby know who is from there and who is not. Unlike in most of the United States, where one can claim to be from a place after a few years of residence, being from Bygdaby means that you were born there. It does not mean being born in one of the small communities located a few kilometers away. Bygdabyingar know who is from Bygdaby and who is not because, first of all, they know many of the local residents, at least by family connections, and, second, because people from Bygdaby speak a dialect of

Figure 1.4
National costumes are worn on special occasions such as Constitution Day, May 17.

Norwegian that is quite regionally specific. Local residents can identify where a speaker is from within up to a few kilometers by their dialect. Shared dialect creates a sense of community, of shared experience, and keenly marks a sense of in-group and out-group.

All of these aspects of life—dialect, community, insiders and outsiders, sense of time and place, tradition—create an intense focus on the local. These "norms of attention," as I discuss in chapter 4, make events that happen elsewhere seem distant and less real when compared with happenings in Bygdaby.

Cultural Homogeneity

A sense of conformity to local traditions, culture, and lifestyle is apparent in the community. Walking along the rural roads on the outskirts of

Bygdaby, one sees well-kept farm buildings, many of which are 100 to 150 years old. Most rural buildings are the same shade of red, white, or yellow. Furthermore, all the houses conform to a similar wooden, square architectural style. There is no variation even on the trim around windows. The houses are almost always freshly painted, and the bright colors make for a cheerful tidiness against blue skies, winter snows, and summer green fields. This sort of cultural homogeneity connects to and reinforces a sense of community, time, and place. Cultural homogeneity is also evident in numerous less visible aspects of life, from information sources to eating habits, clothing fashion, and food products.

Cultural homogeneity has united Norway as a whole in many ways, creating a common community as the entire nation sits down at 7:00 p.m. every night and experiences the same newscasters and the same TV news. Until 1992, there was a single television station—now there are three, and popular satellite dishes are increasingly bringing shows from all over Europe. A high percentage of Norwegians drink milk and eat milk products from the Tine dairy (a farmer-owned cooperative), drink Krone coffee, and eat Freia chocolate. Children throughout Norway have been educated in a standardized school curriculum and learned their grammar and history from the same textbooks across the nation. The same types and brands of California wines are available for the same prices throughout the country in the government-run liquor store, or *Vinmonopolet*. A friend of mine not from Bygdaby told me, "You always know what to talk about if you meet someone; you say something about the TV program that was on last night because everybody saw it. Of course, that isn't as true now that we have more stations."

Despite the greater variety of channels, most Norwegians continue to sit down to watch the same news and television programming at the same times across the country. This cultural homogeneity, reinforced in so many ways, creates a sense of stability and order. This perception of an ordered world provides a great deal of what Anthony Giddens (1991) calls "ontological stability" as well as a rich source of meaning. As Ann Marie, a woman who had lived for many years in California, put it, "When I lived in California, it was wonderful to come back to Bygdaby and feel that there were people and houses around here who have been here for hundreds of years. The farms, the land, the names of the farms, the names of the people. I felt rooted, and I felt that there was a calm and stability, and we had—whether it would be music or food or clothing or traditional festive outfits—it had meaning. And it just felt wonderful." The cultural homogeneity of Bygdaby and, indeed, of Norway in general is a product of tradition on the one hand and of fifty years of considerable government involvement in society on the other.

Although tradition is a salient feature of everyday life throughout Norway, even in the cities, it is particularly important in Bygdaby. Tradition and culture may seem to be about the past, but they hold significance in structuring how people make sense of and respond to present day problems such as climate change. Tradition can be understood, among other things, as storytelling about the continuity of time and culture.

Moral Order and the "Norwegian Sensibility"

There is a feeling of intense order in the streets of Bygdaby. As noted, houses are mostly the same size and style and freshly painted. Cobblestoned streets are cleanly swept. The residents' names and the names of the places where they live are old. These things have apparently been unchanged for a very long time. There seems to be a proper and appropriate way to do everything, from decorating one's home to cooking dinner or going for a hike. Life in Bygdaby has a pattern; it is ordered, known, controlled. As the earlier comment from Ann Marie illustrates, the sense of order provides a strong sense of meaning and community that can be rich and reaffirming.

The order of life in Bygdaby has several dominant themes: sameness, homogeneity, and a certain morality—that is, the sense that there are particular and *proper* ways to do things and that these ways are backed by a moral force. Failure to comply with normal behavior is grounds for suspicion of one's character. As we will see, reactions to climate change are quite uniform among residents and are clearly controlled by social norms of behavior that also reinforce residents' sense of themselves as good people. This sense of moral force is one of the chief ways that the Lutheran Church continues to influence secular society, conveying a sense of the proper way to do things, which I have come to think of as a "Norwegian sensibility" and which, as we will see, is related to Norwegians' strong political and environmental engagement. I think of the Norwegian sensibility as a kind of typical sense of the behavior that is expected from an *ekte Nordmann*, a "real Norwegian." It includes an emphasis on doing things properly and on moral conduct, but also on humility and thrift. One bright sunny afternoon on the deck of the historic hotel where I took a job as a bartender, I met a man in his fifties named Jan. When Jan asked what I was doing in Norway, I explained that I was student and described my interest in "community life." He had lots of thoughts on the matter that he was eager to share. He expressed an almost stereotypical Norwegian sensibility in sharing his love of the mountains and connecting them to leading a simple life: "You know, we don't need money to be happy. All you need is to go up into the mountains on a day like today, maybe with a few friends for community, and that is all you need. It doesn't matter what kind of car you have or what kind of clothing you are wearing; it doesn't make you happy."

Margit, another community member who exemplified the Norwegian sensibility, described her relationship with material possessions:

> One thing that I think about is that I have never been so concerned with material things. The things that my parents really tried to impress upon me was that to have an education, that this was important. But this thing with having and owning so much, that hasn't been something that I was raised to be concerned with. But my mother had good taste; she liked to sew, and she had a good sense of color, such that I have learned about that, and beautiful things make me very happy. I have that pillow that I really love, you know, I have some things that I like to look at. That picture over there of the childhood home of my mother, for example. But I don't buy things other than the things that I need. And I am so happy that my car is functional and so forth.

The Norwegian sensibility is most personified by the generation that experienced World War II and by those who were born just after it. Although there has been a great deal of social and cultural change since then, the Norwegian sensibility remains a standard of proper behavior to which younger generations are still held accountable. The sense of moral order applies to much more than the proper way to keep your house or cook a meal; it is tied to religious understanding about being a good person. In this context, being a good person means contributing to society, holding a strong belief in equality and humanitarianism, and not being wasteful or ostentatious. Many of those who are most active politically around both environmental and human rights issues in Bygdaby connect their social engagement with their religious beliefs. For example, Eirik explained that his Christian faith led him and his wife to their humanitarian work in Africa: "We both felt that to do something, we are a bit involved with the Christian faith, about this thing of caring for your fellow human beings. And the idea of helping the weaker, these were some of the reasons that we wanted to make a contribution for someone who had need for help. It was an important motivating force for us." Per Arne, a man in his early fifties, originally from Bergen, commented, "We have a tradition of equality. We don't have a tradition of showing off when you have wealth, of driving a fancy car. But now we are getting 'new rich.' Now people are showing off their wealth. But this is a recent thing."

This sense of humility, antimaterialism, and thrift is often connected with nature and going hiking or skiing, as evident in Jan's comment and as chapter 5 details. Margit expressed this connection with great feeling:

I like to go skiing in the winter, I go quite often. Sometimes I have the skis in my car, and after work I just put on my boots and go skiing. I always have time to go, always have time to go skiing. And then I don't have to buy new clothes because I stay slim, and I can keep using the same clothing, and I don't have time. I can't go into the stores, use a Saturday, for example. And I think, those who go shopping downtown on Saturday—that is such a terrible thought, to use your time on that. I have all these old clothes, but I can't be bothered to go shopping; I would rather go for a walk.

Egalitarianism is also one of the four characteristics that Norwegian sociologist Thomas Hylland Eriksen (1993) describes as central to national identity. The residents of Bygdaby are clearly concerned about equality. Margit, who spent time with refugees who had been settled in a nearby community, described why she didn't like to buy things:

Margit When I think about these cases where people don't have anything, like in Africa, I think that I should have it even better and nicer, then I would feel that this isn't right. That I would use so many of the earth's resources in order to have such a good life, then it would be even more unequal, I would find it embarrassing. And I see, I have students at school [a Norwegian-language school for refugees]. One of them is coming to visit me tomorrow, but I am dreading it a bit. I think that it is a bit embarrassing for them to see how good I have it.
Kari Because they don't have things so good?
Margit Well, they do really. But maybe they have a little apartment, and they will think, well . . . here there is room for quite a few! You know? I don't think it is so pleasant to have things so much better than others.

The desire for equality and awareness of its absence in the world are deeply troubling for many community members, as evidenced by Margit's statements about the upcoming visit of her less fortunate friend. Students in middle and high schools across the country participate every fall in a series of solidarity actions with people in poorer nations around the world. That November I interviewed a number of young women in their late teens who had organized an event for their school. The six of us discussed the reasons for their involvement in the issue.

Siri   The world is so unequal. It's practically not right . . .
Ingrid (Interrupting, with emphasis.)   It *isn't* right . . .
Siri   . . . to sit here and just get, get, get, that's what we do, at least here in Norway; its totally awful really. Really. It shouldn't be legal.

Or in Margit's words again:

Margit   When my father had just died, and we were going to sell his house in Ålesund, my sister and I, and so right then I had some money, and I thought to myself, I should have given this to someone, but I say to our God [she says looking up, as if asking for permission] that when I die, this will be sold so that someone who needs money can have it, if not my children, unless my children want it. So I can loan it until that time [laughing], yes, that's how it is, you know. . . .
Kari   What kinds of things would you like to do with that money?
Margit   Then I think I would give it to something like SOS Children's Village[1] or some organization like that. That is absolutely what I would do. Because it is so unfair, I have always said it to my students at school, but I must say it to really understand it, that we have been so lucky to have been born here; therefore, we get everything. And those who are born in another place on earth, they are very poor. And [I] think if we had been born in another place and knew that there were others who were born into gold and such, then I would have been angry. I try to take that approach, think about it that way, but it is such a big issue, that I really have a problem myself to imagine it, think if it was us who were there in that other part of the world and were sitting there with nothing. It is just chance.

This desire for equality and the awareness of Norwegian affluence, would surprisingly turn out to be one of the issues that made climate change so difficult to think about, as Eirik noted when I asked him whether he was concerned about climate change:

I see that we do lots of things that most certainly cannot continue. It will work for a while, but sooner or later it isn't going to work. So I am worried in any case for what will happen. And there is so much happening already that is catastrophic for many, but whether we notice it [or not], we have it so good here that we will survive. But there are of course people today, in Bangladesh and other places, who have huge problems and constant problems. So, yes, I am concerned about it.

Moral order and the presence of a moral imperative not only function as a force for social stability in Norway, but also make for the ability to generate moral outrage. Social movement theorist James Jasper describes how one benefit people may experience from social movement participation is the opportunity to voice and refine moral opinions publicly: "It is their ability to provide a moral voice that makes protest activit[ies] so satisfying. They give us an opportunity to plumb our moral sensibilities and convictions, and to articulate and elaborate them" (1997, 5). The moral imperative of the Norwegian sensibility has also been linked to the development of the welfare state and the egalitarianism and humanitarianism at the heart of social policy. Norwegian American scholar Christen Jonassen describes how in Norway

> the political result of the moral imperative was greater equality and concern for the weaker members of society. This demand for social justice found expression in the welfare state. . . . It was the moral imperative that motivated the dramatists, poets and intellectuals of the nineteenth century and the social scientists of the twentieth century to expose areas of social privilege, inequality and injustice. However, if it were not for the presence of the moral imperative in the value system, the lack of justice would not have been perceived. If it were not for the importance of justice in the value system of the Norwegian people, legislative policies and institutions that legitimized and created a humanistic welfare state would not have been possible. (1983, 124)

Norwegians are known for the strong moral stances they have taken on many issues. They give more foreign aid than most other countries and have taken leadership roles on many international human rights and environmental issues. But their very stance on these issues ironically creates an atmosphere in which denial is accepted.

In this chapter, I have described the community of Bygdaby, the ways that residents form a sense of community, their sense of time and place, and the social norms and pressures that they face. I have introduced the concept of a Norwegian sensibility and its relevance to environmental issues. I have described the cultural homogeneity of the community (and of Norway in general) as well as differences between community members in terms of insiders and outsiders and generations. Ethnographic descriptions of how the sense of community in Bygdaby is formed, of local houses and food traditions and the moral order, may at first seem removed from the issue of climate change, but, as we will see, they are inherently linked to responses to this issue.

2
"Experiencing" Global Warming: Troubling Events and Public Silence

We often assume that political activism requires an explanation, while inactivity is the normal state of affairs. But it can be as difficult to ignore a problem as to try to solve it, to curtail feelings of empathy as to extend them. . . . If there is no exit from the political world then political silence must be as active and colorful as a bright summer shadow.
—Nina Eliasoph, *Avoiding Politics*

Passivity and silence may *look* the same as obliviousness, apathy and indifference, but may not be the same at all. We can feel and care intensely, yet remain silent.
—Stanley Cohen, *States of Denial*

Ulrich Beck (1992) argues that we now live in a "risk society," or a society preoccupied with risk. If so, in what sense are people preoccupied? In their daily lives? If people are so preoccupied, why is this concern not visible? If community members consider global warming a significant problem, why does it receive so little serious attention? Are community members simply too busy to do more? In our allegedly rational modern society, we tend to imagine that serious threats or at least potentially serious threats will generate a social response. But this is not the case.

If chapter 1 described the "where" of the story, this chapter describes the "what"—the widespread public silence in the face of the most serious environmental problem we have ever faced. My aim here is to take the reader on a journey in which, to borrow a phrase from Dorothy Smith (1987), "the everyday world [becomes] problematic." Nonresponse to the possibility of climate change may seem "natural" or "self-evident"; from a social movement or social problem perspective, not every potential issue translates into political action. Yet with a closer ethnographic view, we can understand nonresponse as a *social process*. People *might* behave differently. In order to clarify the meaning of nonresponse, this chapter concentrates on the events, issues, and the extent of public silence regarding climate change in Bygdaby. First, I spend some time exploring in depth the contours of public silence in one particular dry and warm winter.

The Events of an Unusual Winter

The way I remember winters, or winters before, you know there was always lots of snow, and it was cold the entire winter.
—Anne, a farmer in her midforties

You know that we have had seasonal changes, but now it has been nearly ten years that we have had a different climate. Less snow and—yes, yes, precipitation hasn't decreased, but it hasn't been cold enough so that it snows in the lower areas.
—Ketil, Culture Center employee in his midfifties

It is, well, milder. There has been less change between the seasons. There is less snow and more like halfway winter, and the summers have been colder. I think that it comes from climate change. Because it didn't used to be this way. And it's, like, known.
—Vigdis, 17-year-old student

Among the unusual local weather events that took place in the fall and winter of my year in Bygdaby, 2000–2001, the most significant and tangible were the very late snowfall and warmer winter temperatures. Record temperatures for the community as reported by the local newspaper in January 2001 showed that the average temperature in the Bygdaby region on the whole was 1.5°C warmer than the 30-year average: October 2000 had an average temperature of 9.8°C (49.6°F), versus the past 30-year average of 5.8°C (42.4°F); the average November temperature was 5.0°C (41°F), versus the past 30-year average of 0.3°C (32.5°F); and December averaged −1.3°C (29.6°F), versus a 30-year average of −3.0°C (26.67.4°F). As of January 2001, the winter of 2000 for Norway on the whole was recorded as the second warmest in the past 130 years. In addition, snowfalls arrived some two months late (mid- to late January as opposed to November). The late opening of the ski area had recreational and economic effects on the community, and the ice on the lake failed to freeze sufficiently to allow ice fishing, once a frequent activity. Ketil described the lack of ice on the lake to me this way:

Ketil Like the lake here, now there is ice on it this year, but until fifteen years ago people came to Bygdaby from eastern Norway, from Hallingdal and other places by train. They stayed overnight at the hotel in order to use the ice. It was completely black out on the ice every single winter.
Kari Because there were so many people, no . . .
Ketil Yes. They went out there and fished. It was very good fishing. But you know it hasn't been like that for the last ten years, now it is completely gone. Nobody comes here any more. Now there is ice of course, but you can't tell if it is safe enough yet, but it will be. Still, you don't see people out there.
Kari I see. So it's less than before . . .
Ketil Oh yes, absolutely. It hasn't been safe ice for nearly ten years now. After a day or two it will rain.

Lene, another community member, also told me about the past ice fishing activity on the lake and how in the previous weekend the local school had to cancel an ice fishing trip for kids because officials weren't sure if the ice would be safe:

When it opened there, they just had two kilometers [of ski run], and not . . . they had artificial snow. And it is a bit [disappointing], with artificial snow. . . . So they tried to do ice fishing, and they tried to have a skating track, but as soon as they were going to have the ice fishing, the weather turned warm. And this weekend they were going to have ice fishing for the kids. It was the trip for the kids. And they had to cancel it because they weren't sure if the ice would hold with many on it, and if they had a lot of kids, it would be dangerous.

The lack of snow in the community was clearly an unusual event, albeit one that people had come to notice over a period of several years.

Community Impacts
The "unusual weather patterns" of the winter of 2000–2001 were, again, visibly identifiable and had tangible effects and measurable economic consequences in Bygdaby. The economic impact on the community from the weather that winter is telling. Late snows delayed the opening of the ski area. The shorter ski season meant that some three dozen employees of the resort (including my husband) began their winter work season in late December or January rather than in November. When the resort did open on December 26, 2000, it was with a single long run of 100 percent artificial snow. No snow had fallen naturally, and aside from the artificial snow on the slope itself, the mountain was bare. Skiers found themselves on a tiny corridor of "snow" between exposed rocky slopes and trees.

The resort owner had to invest 17 million Norwegian kroner (approximately 1.8 million U.S. dollars) in snow-making equipment and another 1.5 million kroner (approximately U.S.$170,000) on electricity and labor to create the artificial snow for that season. The process required the installation of 7 kilometers of water lines earlier that fall, which were used to pump water some 500 meters up the mountainside. In mid-December, it was finally cold enough to start making snow. Thousands of gallons of water were pumped up the hill, and employees worked all night long keeping the snow cannons going. It took a crew of people working around the clock for 14 days to produce snow for the 1,200-meter-long ski run.

This effort testifies to the importance of skiing to the local community, both economically and culturally, and to the fact that unusual weather patterns, whether they were the effects of climate change or not, were very tangible events for community members, including those who owned the ski area, those who worked there, and all those who owned or worked in ski shops, hotels, and in any way with winter tourism. With the ski season shortened, hotels, restaurants, shops—especially ski shops—and taxi drivers were affected.

In addition to economic impacts of the unusual weather, there were also cultural consequences that were not always immediately obvious. On our first night in Bygdaby, my husband and I joined our neighbors Anne and Torstein in their *stabbur*, a traditional storage building for meat and other goods, as they cut up one of their freshly slaughtered sheep. The gathering was a work event, but also a social one. It was the first of several such community events we would attend in which people from the neighboring farms came together to work over a bottle of homemade beer or wine. The six of us crowded around the table in the *stabbur*'s low light and small quarters. Anne worked quickly and skillfully with the knife, instructing her son Lars. A retired butcher, Hans Olav, had come over to help with the project, as had the nearest neighbor, Bjarte, a man in his late sixties. Bjarte brought with him a jug of homemade elderberry wine, which everyone drank with enthusiasm. Sam and I watched this community work event with interest and awe as the slicing and salting of the meat unfolded in very much the same way as it probably had for decades, if not generations. The conversation drifted from sheep preparation and old farm equipment in the building to other topics. Bjarte and his son, it turned out, had recently invented a machine for making artificial snow. The two of them had used it for the first time the previous year in Bygdaby to make artificial snow for the local cross-country ski track. This year their invention would be put to use in several other communities across Norway. "We have had less snow, you know," he said, and after a pause he leaned toward me with a knowing look and a twinkle in his eye, "You know the farmers have to have something to do in the winter."

Bjarte's comment, that we need snow because "farmers need something to do in the winter," alludes to the central cultural significance of skiing for Norwegians in general and for Bygdabyingar in particular. It would be difficult, perhaps impossible, to overemphasize the symbolic and recreational significance of skiing for community members. Norwegians are proverbially said to be born "with skis on their feet." Skiing in this community is the source of a great deal of local pride and identity, and it is a very common pastime during winter months. It is a form of exercise, recreation, and sport. It is not uncommon for children, especially those who grow up on farms, to begin skiing the first winter after they learn to walk.

Symbolic and cultural consequences of the late snow and warm weather are of course harder to measure than economic effects, but they are equally real. Skiing is a very old, traditional activity. It is "what farmers do in the winter." Cold snowy winters, required for skiing, are expected in Norway. The arrival of cold temperatures and snow marks the changing of the season, part of the sense of the ordering of time and place that forms the sense of moral order and ontological security (see chapter 1). Norwegians desire, as the phrase goes, a "sikkelig Norsk vinter"—a "real (or proper) Norwegian winter," one that gets really cold and forces people to wax skis with *blå smøring*, blue wax, for cold temperatures. The phrase "sikkelig Norsk vinter" suggests the deep cultural significance of mountains, skiing, and snow to the Bygdabyingar's identity, to their concept of "the good life" and what is normal and right. The winter season is marked by the ritual of skiing. Participating in this activity assures community members of where they are in space (at home, in the mountains, in Bygdaby) and time (you know that it is winter when you put your skis on). Skiing as a ritual connects community members to the place where they live.

National and International Events and Media Coverage
The local weather wasn't the only thing that prompted thoughts about climate change that winter season of 2000–2001. On the national and international level, October and November brought extensive flooding across western Norway. During the same time period, communities

Figure 2.1
Climate change in the news: Headline reads "Bad weather will give you higher insurance."

across England received record storms that led to their own dramatic flooding. The topic of climate change was clearly salient in the media, and information was widely available.

In addition to local weather events that provided the possibility of direct experience of climate change that year, there were a number of national and international political events that brought the issue to Bygdabyingar's minds. Just three months before my observations began, the Labor Party had replaced a coalition government that was described as "the first government to fall due to global warming." The former prime minister, Kjell Magne Bondevik, had opposed the construction of natural gas facilities on the grounds that they would generate too much carbon dioxide (the primary gas associated with global warming). Bondevik had taken a stand against the plants, called for a vote of confidence, which he lost, and then resigned. In November, several hundred miles to the south, the nations of the world held climate meetings at The Hague. And from January to March 2001, three events focusing on climate change received significant attention in the national Norwegian press, making the front page of the national newspapers that were delivered on doorsteps and for sale on racks across Bygdaby.

First, the New Year's Day speech by then prime minister Jens Stoltenberg raised the issue of climate change, connected it with a reference to the unusual weather of the previous fall months, acknowledged the failed climate talks, and urged people to take individual and collective responsibility:

> We have had strange weather recently. In the south and east parts of the country, we have scarcely seen the sun, and we have had flooding and precipitation records. Farther west and north things have been dry. In Sunndalsøra, it was 18°C in December, and people could pick roses in the middle of advent time. An unusual winter isn't proof of climate change. But change over time tells us that we must take this seriously. It was a defeat that the climate meetings at The Hague fell apart. But it tells us also how difficult it is for the nations of the world to come to agreement on sharing the burdens. I am convinced that it is possible to come to agreement, and we are doing all we can to contribute. To meet the climate challenge will require contributions from all of us, as a society, businesses and individuals, but this is the way forward. (Nationally televised speech, January 1, 2001, my translation)

Another New Year's Day speech, given by King Harald, also mentioned climate change, although this time more briefly and in conjunction with an emphasis on Norway's relationship with the United Nations: "We stand together to protect the peace, to forbid weapons that kill arbitrarily and bring war criminals to court. We work together to fight

AIDS and other epidemics, to watch over climate change, and to make clean air and water available for all" (nationally televised speech, January 1, 2001, my translation).

The fact that both the king and prime minister chose to address the issue of climate change in their New Year's Day speeches highlights the salience of this issue on the national level. By mentioning global warming in this context, both leaders called on the nation's citizens to keep this issue in their minds as significant and relevant. Yet although these events were visible and public, they portrayed climate change as an abstract concept. These speeches were about large, significant issues. They served to call attention to global warming, but doing so did not automatically make the issue "real" or integrated in daily life.

Second, three weeks after these speeches, on January 22, 2001, the United Nations Intergovernmental Panel on Climate Change released its third assessment report on climate change. This report disclosed that climate change appeared to be occurring faster than before (average temperatures may rise 10.5°F in the coming century as compared with previous estimates of 2.5 to 5.5°F) and gave strong evidence of human causes of climate change. The findings of this report were unanimously approved by some 150 scientists and 80 members of industry and environmental groups. The next day, on January 23, the front page of the *Bergens Tidende (Bergen Times)* featured a large picture of a walrus sitting on a melting iceberg with the headline "Varmere, varmere, varmere" (Warmer, Warmer, Warmer).

And third, in March climate change made headlines again as President George W. Bush announced that he was pulling the United States out of the Kyoto Protocol because it was "not in our economic interests." I first learned this news not because I read the *New York Times* on the Internet or even because I watched the evening news. I learned of it because Bygdabyingar were eager to tell me and discuss it. Conversations that condemned the United States were easy to come by in Bygdaby (see chapter 5).

Thus, the national news media *did* discuss climate change in connection with coverage of the weather and described future weather scenarios and impacts—in contrast to the local news, which I discuss later. In the same time period that unusual weather patters were observable in Bygdaby (November 1, 2000, through January 31, 2001), there were 34 stories on climate change in the largest papers in Norway (*VG*, *Dagbladet*, *Aftenposten*, and *Bergens Tidene*). Issues associated with climate change made the front page of these papers multiple times during that period. In some cases, these stories were accompanied with dramatic photos (see, for example, figure 2.2). Newspaper coverage generally portrayed climate change in a political way. National press provided information about climate change and made connections to weather events. So what happened to this information once it entered the community?

The Contours of Public Silence

I know that environmental protection is very important, but in the everyday I don't think so much about it.
—Lisbet, early forties

As a sociologist, I was as puzzled by the behavior of the people I met as I was concerned about the lack of snow. It is relatively difficult to imagine and describe behavior that is missing, something that might have occurred but did not. This approach is related to the study of everyday life, of how, as Dorothy Smith (1987) has explained, the everyday world that we see as normal is in fact "problematic." To see the production of "normal reality" as a social process requires one to pay careful attention to moments when topics appear briefly, under the surface of an interaction, only to disappear again, as the topic of climate change did in Bygdaby. I look more closely now at the spaces where we might expect to find some response from local community members: formal politics, volunteer associations, street protest, local media coverage, and daily conversation.

Political Organizing
The most obvious way of responding to global warming's significant threat to society, culture, and the economy, especially because it is such a large problem, is through some sort of collective political action. Political response may manifest in a number of ways. Official actors may enact and carry out policies through the formal political system, and citizens propose ballot measures, vote, and testify at public forums. Political response also occurs as volunteer groups organize, raise money, put pressure on official channels, and write letters to the editor. And political response happens as people take to the streets in direct protest. Each of these avenues of political action is animated when people get together and talk about issues in their lives. Political philosophers such as Hannah Arendt and Jürgen Habermas describe the central importance of public

Figure 2.2
Climate change in the news: Headline reads "Warmer, warmer, warmer."

dialog to the generation of power and the project of democracy. It matters, these philosophers tell us, whether people get together and talk about their lives. If, as Arendt famously tells us, "powerlessness comes from being inattentively caught in the web of human relationships" (1958, 183), then political talk is a key antidote. Political talk is a form of meaning-making power; it engages the "sociological imagination," that "quality of mind necessary to grasp the constant interplay between our private lives and the political world" extolled by sociologist C. W. Mills (1959, 13). We turn first to the more visible, obvious dimensions of political activity in a community.

Formal Politics: Climate Change Is Not a Local Issue   The level of direct participation in the political process is quite high in Norway, and Bygdaby certainly follows this pattern. At the time I lived in Bygdaby, at least eight political parties were visibly active in the community. Community politics are structured in such a way that some two dozen members sit on the town council, the *kommunestyre*. In addition, each party maintains a "list" of potential council members. When election time comes around, the number of people on the list who take seats on the *kommunestyre* depends on the number of votes the party receives from the community in that election. Thus, in addition to the several dozen community members who sit on the council itself, another half-dozen or more might be on the list (i.e., reserve duty). And behind those people are party members who participate in the regular party meetings at which the party's agenda and local politics are discussed. Thus, for a town of 10,000 residents, there is a great deal of direct citizen involvement in the local political system, especially in comparison to the low political participation in the United States.

So what went on in these meetings in 2000–2001? Was climate change discussed? Meetings focused on a wide variety of topics from zoning laws regarding downtown to new area for huts and, of course, the budget. To my surprise, climate change was never discussed in any of the city council or specific Labor Party strategy meetings I attended or even in the meetings of the municipal subgroup on culture and environment. If people actually cared about the issue, how could this be? It seemed that from the perspective of both the newspapers and the local politicians, climate change was not a local issue at the time. When I asked Øystein, a member of the Labor Party who sat on the city council and was the father of several preteen children, about climate change, he responded: "Yes, it is of course a theme that everyone is interested in, but locally it isn't discussed much because, well climate change, you know there isn't so much you can do with it on a local level. But of course everyone sees that something must be done in order for, in order to stop the development of climate."

Øystein referred to the sense that climate change is more of a national issue. From an organizational standpoint, climate change is clearly a global problem, and nations must work together. Nonetheless, it is on the local level that people in Bygdaby experience changes, that people's lives are affected by cultural or economic impacts. Labeling climate change as a national or international issue and "not a local one," although understandable, also works as a justification for the public silence. Susan Opotow and Leah Weiss, in their review of denial with respect to environmental problems, describe the displacement of responsibility onto "higher authorities or legitimate decision makers" as a form of denial of self-involvement (2000, 481).

Volunteer Organizing Volunteer organizations are widespread across both Norway and the United States. As people come together in organizations from parent–teacher associations to church groups, the Red Cross, and local chapters of larger nonprofits, they form the basis of a public sphere. Participants in these groups not only achieve the organization's manifest goals, but also come to see themselves as part of a larger whole and develop collective interpretations of the world. For Hannah Arendt (1958, 1972), both a sense of collective vision and identity are key elements for citizenship, which is in turn key for democracy. Jürgen Habermas (1984) has also been highly concerned with democracy. His theory of communicative action describes the importance for the lived experiences of people (their "life world") to inform social institutions and the society as a whole. For Habermas, the fact that volunteer organizations are controlled by neither the state nor the market makes them potentially important sites where people can get together, share their experiences, and from this sharing develop their own narratives about the causes, consequences, and potential solutions to problems in their lives. These narratives can then form the basis of social movements. Furthermore, through such interactions in voluntary organizations, people may develop appropriate ethics for citizenship.

Thus, a second area of local political influence to consider is that of volunteer organizations. If political participation was high in this Norwegian town at the beginning of the new millennium, volunteering was off the charts. Community members were actively volunteering in everything from Amnesty International and international children's hunger relief groups to organizations set up in opposition to the European Union. They also participated in community discussion groups such as the Saturday Academy, a community group that met roughly once a month to discuss local issues of interest such as a proposed shopping center to be located just outside the downtown. Yet there was no active environmental organization, and in none of the dozens of local meetings I attended the year I lived in Bygdaby did climate change come up, not even once.

In the Streets Political activity happens not only in the formal political system, but also in the streets. Political activity in the streets is activity of protest. Like volunteer organizations, the public protest in the streets can form a vital piece of the public sphere. Residents have the opportunity to share their views and to hear what others have to say under circumstances that, although often still scripted, are generally more free form than within the meetings of volunteer organizations or political parties. Here, too, is an important space for collective meaning-making on the issues a community finds most pressing. During my eight-month stay in Bygdaby, residents took to the streets with signs and songs on several occasions. The first event occurred shortly after I arrived, on November 9, the so-called Crystal Night on which the Nazis seized many Jews and took them to concentration camps.[1] A group of 30 or more young people met at the bridge. After several short speeches describing the importance of the date and the continuing presence of racism in Norway, the group walked through the downtown carrying signs and candles and singing. Most of the participants were in their early twenties, although there were also a few people in their thirties, forties, and fifties.

All in all during that year, climate change was mentioned, as far as I'm aware, only once in a public forum, as part of a highly motivational speech given by a woman in her late twenties in a downtown public event on May 1:

And I am seriously concerned and afraid of the greenhouse effect; we can see that flooding and deserts are coming. So I wonder what it will mean for life on Earth and for the humans who live here. A Swedish friend of mine said that we have to cut our CO_2 emissions by 60 to 80 percent. The Kyoto Agreement is about cutting them by 5 percent. And even that ridiculous pace was too much for the climate hooligan George W. Bush in the United States. The head of the USA's Environmental Protection Agency said that "we have no interest in meeting the conditions of the agreement." Well, that may be so. But it is other countries that will be hit the hardest from climate change. . . . But now it isn't that it's just Bush and his cronies who sin against the environment. The government of the Labor Party has said that so long as the United States doesn't do anything with the climate agreement, so in a way it is just fine for the rest of us to shirk our duties. At the same time, they are putting on the pressure to build pollution power plants. As a result, Norwegian emissions are going to increase even more. And that is of course the opposite of environmental protection! It should really be that if the USA begins to break environmental promises that were agreed upon in Kyoto, it is just natural that Norway should take the climate treaty seriously and follow up on it. (my transcription and translation)

The references to the United States construe Bush policy as a bad example that Norway should not follow (I discuss the use of criticism of the United States as a rhetorical strategy in chapter 5).

All in all, *Bygdabyingar* were remarkably active politically in this period, through the formal political system, volunteer organizations, and street protests. They actively engaged with issues ranging from opposition to the European Union and racism to land zoning and the local labor market. But in each of these spaces climate change was virtually invisible.

Local Media Coverage
Newspaper stories are considered not only a key source of information, but a central element in how the public "frames" information. In Bygdaby, the local newspaper, the *Bygdaby Posten*, is an especially key source of information about the tangible, everyday situations that William Gamson (1992) describes as a necessary complement to abstract information so that people can generate collective action frames. Especially for a community that is very self-conscious, events reported in the local paper—i.e., the things that happen down the road or that affect one's neighbor or cousin—are more likely to seem real.

On Tuesdays, Thursdays, and Saturdays the local paper comes out in Bygdaby. The 10- to 12-page paper enjoys close to 100 percent readership in the community. It covers local events, with occasional mention of related regional and national events, and features color on the front and back pages. It also has a sports section (dominated in the winter by skiing events), an editorial page, and a lengthy classifieds section that contains remarkably detailed information on local events, from upcoming organization meetings to special events. A reader can even learn if her dentist will be taking a holiday soon by reading this section of the paper. Also included in the paper are television and radio programming schedules for the three-day period. Each issue has two to four letters to the editor. The Saturday edition features a local resident and an historic photo of some detail of life in Bygdaby from the past century. The front page normally has a large photo for the feature story and seven or eight smaller stories. The three-day weather forecast also appears on the front page, along with a highlighted quotation from someone in the community linking to an inside story. The back page generally features a photo essay in the "human interest" genre.

So how did the problem of climate change play out in the local paper in 2000–2001? Did the newspaper help people to make sense of their troubling concerns? Did it help them to develop a sociological imagination—that ability to connect local experiences to wider events? During the three-month period from November 2000 to January 2001 when the weather was most unusual in Bygdaby, I found 11 articles that made references to the lack of snow or to unusual weather. Most mentions of the winter's weather patterns in the local paper were made in connection to skiing and the ski area's efforts to create artificial snow. So there was coverage of the unusual weather. However, as we shall see, these stories talked about the weather in ways that were subtly reassuring. In mid-November 2000, a large spread of photos and text described the new snow cannons and efforts that would go into the making of snow. This story had a reassuring quality; snows were late, but with the wonder of technology the ski season would be saved, and all would be well. Perhaps this tone implied an effort to avoid scaring off tourists. The manager of the ski area was certainly carefully positive in his statements about the good quality of the snow. A story at the end of November described how one local ski club had to travel several hours by train to practice at another location for the first time ("Ski Talent Got to Test the Snow"). This article did not mention why the team had to travel, nor did it raise the issue of climate change or speculate on what might happen in future years. Nor was there any mention of the impacts on the local economy, when local ski groups traveled elsewhere for training. Rather, the writing framed such happenings as unique events. Instead of drawing attention to the possibility of climate change or raising questions about local weather in light of scientific predictions, these news stories worked to reassure readers that "all is well." Other stories in the sports section referred to the condition and quality of the artificial snow at the cross-country track in connection with races. A later article (January 4, 2001) with the headline "Disappointingly Few Take Advantage of the Artificial Snow at Bavallen" described the manager's frustration with the lack of business at the ski area. On January 9, a large story in the sports section (page 6) recounted that more than 200 individuals raced in the local cross-country race and that the race took place on 100 percent artificial snow. The last paragraph of the story reported that none of the racers complained about the track. One racer was quoted as saying, "We took the conditions as they were, and all went fine." On January 11, the headline of a column on ski conditions on page 3 included more encouragement in the face of the snow-free slopes: "The Ski Slopes Are Waiting," and a subhead stated, "It may be that the chances for a weekend ski trip are improving."

Because the weather report appeared only in a small box on the front page of each issue, the paper did not contain a "weather section" per se, in which the issue of climate change might have been taken up. The closest the weather came to making headlines was the front-page story on the day it became cold enough to begin using the snow cannons at the resort. The paper for December 19, 2000, featured an extra large front-page spread with two photos of the snow cannons in operation. The headline read: "Historisk Snø i Skianlegg" (Historic Snow in Ski Area). Subheadings read: "Finally it is snowing in Bavallen. Sunday it was cold enough for the new snow cannons to do their work, and since then the production of artificial snow has been more or less continual." The remainder of the story gave a hopeful description of the number of snow cannons in operation, the expected opening date of the ski area, and what turned out to be an overly optimistic description of the expected ski conditions.

Two stories covered the warmer weather specifically. On November 4, 2000, page 2 featured a story with the headline "Mildest October in 40 Years." This short article reported average temperatures as recorded at the local weather station and compared them with past years. There was no mention of climate change or analysis of what might be happening. A second and much longer story appeared on January 6, 2001, on page 3 with the headline "The Year 2000 Was Warm and Wet." This story gave a number of statistics for the months of October, November, and December from the local weather station and compared them with national-level data. Again, no reference to or discussion of climate change appeared.

A third story related to unusual weather was accompanied by a large color photo on the back page on December 5, 2000. The photo featured a giant orange chanterelle mushroom in full color. The headline read: "Advent's Chanterelle" with a large subheading: "Completely impossible, thought Stig Dalgard as he, on a trip to his mountain farm this weekend, came across something that could only be a chanterelle." The article continued: "Stig Dalgard has found mushrooms by his farm for many years but never so large as this. And for certain not so late in the year. When the Christmas month comes, it isn't usually mushroom picking that one is up to when one goes to the mountains. But for Sunday dinner on the first Advent day Stig will be able to serve freshly picked mushrooms." The article continued with discussion of how he came across the mushrooms and concluded with the statement that "it is usually a skiing trip, not mushroom picking, that brings people to the mountains at this time of year. But many things seem to be against the norm this fall and winter."

The color photo of the bright orange mushroom and the apprehensive tone of the reporting made this story different from the others. This news story was one of the few that did not provide a sense of reassurance. Although comments about the mushroom and serving mushrooms for dinner may seem to trivialize this event, in fact the mention of Advent, mushroom walks, and ski trips connected this event to deeply significant social categories for community members. The impact of this short story resulted from the way that it drew on Bygdabyingar's knowledge of the local nature and their sense of time and place. Unless people are aware of the natural world and thus attuned to the "normal" patterns of natural phenomenon—when mushrooms fruit, the seasonal timing of snowfall—environmental change will go unnoticed. The native people in the polar regions, who have reported "on the ground" effects of climate change, have done so because they are aware of "normal" patterns and cycles. Bygdabyingar also have this kind of perception of their region—it is brought into collective awareness through ritual practices of autumn mushroom hikes and winter ski trips. To find a large chanterelle in December during Advent interrupts the sense of time and place, going against the flow of order and tradition, which for Bygdabyingar are connected to the land and its seasonal cycles. This chanterelle story was different from other news coverage of weather-related events in that, rather than reassuring readers that all was well, it was one of the only stories that explicitly opened the door for questions such as "What is going on? What does this mean? What will happen next?"

Finally, the lack of snow that year was the subject of a playful advertisement in which a local member of the ski community stood holding a pair of skis on top of the local newspaper on a grassy slope with Bygdaby in the background. The ad was for advertising in the local paper. The text of the ad read: "Without the *Bygdaby Post* I'm just standing on bare ground." As I described in chapter 1, the use of clever jokes is a kind of traditional practice for Bygdabyingar. Jokes and humor play many social functions, and making light of serious issues can be a way of releasing the pressure associated with the uncomfortable emotions raised by taboo topics.

In the most political space of the paper, the editorial page, I found no editorials or letters to the editor commenting on the weather or making political links between the poor weather and the economy. There was no reporting on economic impacts of the shorter winter season for the ski area or the community, nor on the economic success or failure of the winter season for the community that year in general. There was also no reporting about potential future consequences that climate change might bring should trends from that year continue. In fact, the term *climate change* did not appear in the newspaper at all that season—but, interestingly, not because the chief editor, Lene, did not herself believe climate change was happening. In an interview held in late December 2000, we discussed the issue at length:

Kari One thing I have wondered about since I arrived, that is that the snow came late and that there isn't as much as there has been.
Lene For tourism and that kind of thing, is that what you are thinking of? Or that the snow is away?
Kari Yeah, and do you think—that it is less than before?
Lene Oh yes. Yes! Both snow and cold temperatures.
Kari And cold temperatures?
Lene Yes, both. And of course it's climate change that is doing it. There isn't anything to be done about it. And we are on our side of the mountain—ten years ago the weather came from the west, so the snow came down here. Now it comes from the east, and the snow falls on the east side of the mountain.

Lene's perception regarding links between observable weather and the possibility of climate change was not translated into the reporting in her paper, however, even though the dry and warm weather was clearly troublesome to her. When, with skis in hand, I ran into Lene outside the train station a month after we met, it was snowing hard. After we greeted one another, she noted with satisfaction, "Now you have yourself a real Norwegian winter."

On the whole, the local newspaper in Bygdaby conveyed information about unusual weather events, yet with the exception of the story on the chanterelle, it did so in a normalizing way that created a sense that "all is well." Stories alluded to the problem only to show that things were in fact going fine, emphasized the wonder of technology, and later complained that there were so few skiers "taking advantage" of the artificial snow—as though the problem was with the skiers, not with the lack of snow! Local stories about these weather events could have made links to similar events across Norway or the world. Columnists could have asked questions about the long term consequences of the situation or reported on the poor quality of artificial snow for skiing. Yet they did not. Only by reading between the lines would one get a sense that something big may have been taking place that year. News stories covered the installation of new snow cannons, the annual winter ski races, and the local ski organizations' latest activities. These stories were pitched to answer the questions that everyone was asking: "What will happen now?" "How are things going for the community this year?" The stories addressed these questions repeatedly and with full-page photo spreads. Yet at the same time that they conveyed specific information about atypical events, they smoothed over troublesome questions with references to technology, appeals to Norwegian toughness and self-sufficiency, and the use of humor. The story on snow cannons emphasized the wonder of technology; the ski racers "took the snow conditions as they were, and all went well." And the newspaper used humor about the lack of snow as a clever means of advertising.

Conveying troubling information yet doing so in a reassuring way is achieved not only by emphasizing the power of technology or spreading the "take it as it comes" variant of Norwegian "toughness," but also by following local norms about a sense of place and time (chapter 1). Indeed, one might say that the newspaper is a key site for reproducing such norms.

Among the important threads of scholarship upon which I have drawn is sociologist Nina Eliasoph's very provocative work on apathy in the United States. In her study, Eliasoph found that local news presentation contributed to the lack of sociological imagination for the people she met by providing local stories without explanations of causes beyond local circumstances: "the local news made it seem that politics happens elsewhere." Eliasoph concluded that "reading the local newspapers did not help citizens make connections between politics and everyday life" (1998, 226, 210). The same can be said of the local paper in Bygdaby. The tendency to focus one's attention on the local in 2000–2001 made it less likely that residents might make links between their weather and information they learned about polar ice melting and other effects of climate change. Information connected to climate change might be "known," but residents easily kept it in separate mental categories from what was happening around them in daily life. The mental orientation toward the past, reflected in the frequent historical photographs in the newspaper and in long-time residents' experience of weather changes, increased community members' sense that something unusual was going on, giving them ready access to a mental "data set" of past winters. I believe that this awareness of the past contributed to so many people's suspicions in Bygdaby that the unusual weather *might* reflect climate change. Their consciousness of the past gave them a frame of reference that made the problem visible. Yet without a corresponding focus of attention toward the future or across spaces so that other unusual weather taking place that season came into view, the implications of present events were hard to understand. Information was known in an abstract sense but not integrated into the sense of immediate reality.[2]

No Place to Talk about It: Conversations about Global Warming

In early January, it had finally snowed enough in the high country to ski in the Bygdaby area. I was on my first Sondagstur or "Sunday Trip" out on skis with the local hiking and skiing club, Bygdaby Utferdslag. The sky was clear, a beautiful very pale light blue unique to the low-angled winter sunlight. We spent the morning climbing up a steep road to a small saddle where we stopped for lunch. On the way back, a man in his fifties skied alongside me. He wanted to know where in the United States I was from and asked some questions about Washington and Seattle. After a bit, I asked him about the ski route into one of the nearby huts that I had seen on the map. He told me where the route was but said that there wasn't enough snow yet and that the route was still pretty rocky. "You know, there is less snow this year than usual." I nodded but said nothing and waited to see what he would say next. He added, "You know that they have been talking about climate change and that it is the fault of humans." I asked him what he thought about that. He gave a funny laugh and then said, "The United States has been reluctant to decrease their emissions." "Yes," I agreed, "That is terrible." At this point, there was a long pause as we skied side by side. The topic was a grim one. What more was there to say? We continued skiing in silence.

—Field notes, January 6, 2001

Before an issue can make it into a council meeting, onto picket signs, into the framing of a local news story, or into a newspaper editorial, somebody has to start talking about it. When people get together and talk, a number of important things can happen. Conversation is the site for exchange of information and ideas, for human contact, and for the building of community. Conversation can help people understand their relationships to the larger world or obscure them. It can engage the sociological imagination, that "quality of mind necessary to grasp the constant interplay between our private lives and the political world" (Mills 1959, 13). Conversation can also do the complete opposite. Specifically political talk is important because, as Eliasoph tells us, "it helps cultivate a sense of community so that people care more, and think more, about the wider world" (1998, 11). By "political talk," Eliasoph means the kind of conversations through which people might come to see local phenomenon within a larger context, to see "personal" events as "political." She draws upon the work of Hannah Arendt, who describes political talk as an end in itself because it is a means of creating one's own interpretations of reality, a necessary form of political power. For Arendt, "power springs up between men when they act together," and powerlessness comes from being "inattentively caught in the web of human relationships" (1972, 151). Through engaging in political talk, people can figure out what they think, cultivate concern for the wider world, develop a sociological imagination, and participate in the form of political power that comes from having their own interpretation of the world.

Building on Arendt's observations about the power of citizens' political talk in the shaping of their point of view and of their awareness of what issues are on the political agenda, Eliasoph writes, "Power works in part by robbing the powerless of the inclination or ability to develop their own interpretations of political issues. With active, mindful political participation, we weave reality and a place for ourselves within it. A crucial dimension of power is the power to create the contexts of public life itself. This is the power to create the public itself." She goes on to note that "this power can also be a means to more instrumental kinds of power because it opens up some aspects of life for public questioning and closes off others, allowing some aspects to seem humanly created and changeable and others to seem natural and unmovable" (1998, 17).

Eliasoph further notes that it is not just the topics that people must be able to control in order to be powerful, but the ways of engaging: "Without this power to create the etiquette for political participation, citizens are powerless. Without the power to determine what sorts of questions are worth discussing in public, citizens are deprived of an important power, the power to define what is worthy of public debate, what is important" (1998, 17–18). Yet it was precisely this kind of talk that was in limited supply in Bygdaby at the beginning of this century. In the words of Mette, an 18-year-old student: "Well, yes, we talk about it, especially *this* year, if you are sitting with someone you know, you talk about the weather. But we don't talk about it in a *serious* way."

Throughout the year I lived in Norway, climate change was frequently present in casual conversations. For example, in late December 2000, an elderly woman getting onto the bus commented to the driver that "this climate change seems to be a good thing." She presumably felt this because it was easier for her to walk around without so much ice. Yet I noticed that people talked about climate change in a particular way. Discussions of climate change were associated with a significant degree of concern and with feelings of helplessness and powerlessness. The issue was raised with a shaking of heads, a sense of trouble and discomfort, with no notion of what could be done, as when the skier in my section-opening story mentioned climate change to me in relation to the lack of snow, but then the conversation stopped. I also noticed that there were places where people did not talk about climate change. Local political leaders mentioned climate change more as an inspirational topic to try to motivate action, but not as a day-to-day concern.

From my observations, climate change was most likely to be discussed as a serious topic when it came up on the news and there were several people in the room. I did not have access to many of the places where such conversations might occur, such as at home between friends or with the family around the television. Because I could not be present for all these conversations, I could learn only indirectly about what might have been said by asking residents during interviews. I made a point of asking people whether they were talking about these things privately. Here is the reaction of a group of young female students:

Kari Do you tend to talk about climate change or the future with your friends, or . . .
Siri No, but it has been a bit . . . like, this year that plant there shouldn't be flowering, like (pointing out the window). . . . In fact, there should be snow here, so you know this is totally crazy.
Trudi You know when it has come as far as the first of December and there isn't snow . . .
Kari So you do talk about it?
Siri Well, not seriously, though.
Trudi Not seriously, but it is usual in a way. You know, when you watch the news, you might talk about it.

These young women were particularly politicized, actively following the news and even working in their school to educate their peers about

AIDS. The fact that even they discussed climate change only "a little" further suggests that such conversations were not widespread. When I asked Øystein about whether he talked about climate change with his children, he told me that he didn't and referred to the fact that the kids get it in school.

Kari Is it something that people, well, do you talk about it with your children?
Øystein Not so much, more sporadically. It isn't often, but we have of course; of course they have this theme at school, the kids. They have it at school, of course, so they hear it, of course, they are aware of it. They have it at school so that the kids will have a good relationship with the environment and so forth, but I don't talk so very much with my kids about it. Not so very much.
Kari No?
Øystein No, I don't, very little really. We don't talk about it much.

Yet the educators I spoke with in Bygdaby expressed significant difficulty speaking about environmental problems and climate change with students. They had a hard time connecting the issue to students' lives and struggled with the sense that this information was somehow "too much."

Most telling was the fact that when I brought up the issue of climate change in interviews, I noticed that it often killed the conversation. People gave an initial reaction of concern, and then we hit a dead zone where there was suddenly not much to be said, "nothing to talk about." From the things they said (e.g., "it's depressing," "not sure what to do"), I had the distinct sense that it was an issue that people were uncomfortable discussing. It was not, for example, a topic on which the conversation flowed, or when the topic was raised, people gave "stock" replies, easy answers, or even easy reflections. These reactions were similar to those that John Immerwahr observed in focus groups on climate change in the United States: "As they thought about the problem, they seemed to run into brick walls, characterized by lack of clear knowledge, seemingly irreversible causes, and a problem with no real solution. As a result they were frustrated and eager for a solution but unsure of which way to go. The symptoms of this frustration are clear. The first is that people literally don't like to think or talk about the subject. Our respondents always seemed to want to move the topic from global warming itself to more familiar topics, such as moral deterioration, where at least they felt on firmer ground" (Immerwahr 1999).

For example, in a joint interview with Åse and Lisbet, two women in Bygdaby who worked on human rights issues, the conversation did not stay focused on the problem of climate change. Instead, the women turned the conversation to the more manageable topic of electricity prices and the fact that the power companies were earning so much money. Neither of them answered my question of whether they ever spoke about climate change with their friends:

Kari I wonder if this [climate change] is something that you think about or worry about or talk about with your friends?
Åse I don't know, but I try to influence the government, you can say. Because they are the ones that are going against all the agreements, so they are the ones that are trying to lift the brakes.
Lisbet It's like I think that everybody knows about it, but just what they will do, concretely, that's not so easy to figure out.
Kari What they will do, or what you will do?
Åse What they will do isn't so easy to figure out. But people know about it, I think so. And some people feel that's just the way it is.
Kari Yeah.
Åse It's just a joke to get involved with that.
Lisbet Of course, we are used to it . . . just look at the gas prices or electric prices. It's like people are trying to lose by building power plants.
Åse But the ones who are making money on it are the power plants.
Kari Because prices are going up.
Åse Yes, we pay nearly twice as much in electricity. . . . Yes, Norway has of course a lot of energy, and that is, you know, a great way to keep the prices high, and then we accept the construction of new plants—they're making new ones here and there. . . .
Lisbet In the everyday, I don't think about it so much, but I know that environmental protection is very important.
Åse Yeah, you say in the everyday, what . . . well . . .
Lisbet Yes, climate change, you know we have had that since the ice age, so there has been climate change all the time.
Åse Yes, it happens all the time.
Lisbet So I don't know how much more the atmosphere can take. But I don't think so much about it. I'm worried enough, but . . .
Åse No, I don't think about it myself. But when the government does something, they do it completely. I think that most people individually do what they can, when it comes to saving electricity and all that here. But what I think is so bad when they increase electricity prices is that it is the power companies that are making money off of it.

When I asked people if they talked about climate change with their friends, the answer was generally an awkward and almost apologetic "no." Lisbet, like others, said that she was worried but didn't think much about it in her everyday life.

People in Bygdaby did not describe economic losses that might be tied to climate change—neither present nor potential future impacts—in terms that would lead to political analysis (e.g., blaming industry, the Norwegian government, or the United States for disproportionate emissions). What did happen when I raised the issue of climate change was that people indicated that they believed it was happening, that it was a serious problem, and they often gave quite explicit information about the history of local weather events in their community. Although I did not ask specific knowledge questions (such as whether people knew the causes of climate change or knew how much Norway was contributing to overall global emissions or knew the largest sources of emissions in their community), rarely did I get the sense that people were ill informed on this topic. Rather, the topic was one in which knowledge existed in "raw form." It had not been processed by political analysis. People knew basic information, but they hadn't done much with the knowledge. They had not developed the nuanced understanding or the free and easy speech patterns that result from active participation in political discussion of a topic. In fact, I always felt that raising the issue in interviews was considered a bit "off topic." It made people uncomfortable; it was somehow not as polite as the questions I asked about local traditions or what people liked best about their community. Instead, as I discuss in chapters 3 and 4, the issue of global warming was one that had to be carefully managed. It brought up potential feelings of helplessness and despair. As all the educators in Bygdaby stressed, there was always a need to be positive when discussing these issues.

Bygdaby community members did respond directly to changes caused by global warming in some ways. Bjarte and his son had invented a snow-making machine for the cross-country track, and the owner of the ski area spent a great deal of money to purchase snow cannons and to make artificial snow. As discussed earlier, casual comments about the unusual weather were commonplace and in general contained references to climate change. And there were economic responses in the community. Yet there was a general failure in the community to respond in any political way, either through social movement activity or through the kinds of conversations that might generate such activity. There was no local political organizing; few, if any, conversations took place in which community members talked about what could be done, what might happen next, and how to plan for it; and no one wrote letters about the topic to the local paper.

Instead, information about global warming remained outside the sphere of normal life, normal thought, and the sense of what was normal reality. People were aware that there was the potential for climate change to alter life radically within the next decades, and when they thought about it, they felt worried, yet they did not go about their days wondering what things would be like for their children, how farming might change in Bygdaby, or whether their grandchildren would be able to ski on real snow. They spent their days thinking about more local, manageable topics. Ingrid described how "you have the knowledge, but you live in a completely different world." In the words of one environmentally active man in his midforties,

I don't think we can get around feeling problems by pushing it out and pretending it isn't there. Everyone says that everything is so sad and sorry that they don't want to hear about environmental problems. But they know. They know that there are serious problems. . . . And I must say that I don't know anyone who goes around and is bothered in their daily life due to environmental problems. I don't do it, and I don't know anyone else who is. But that in between it is discouraging and an emotional weight I don't think that can be avoided.

Although present in small talk, climate change was not a topic of the more serious political talk that took place within the community's many volunteer and political organizations. It was not, for example, discussed at any of the multiple meetings I attended of the local Labor Party branch. These meetings were centered on very "local" issues, such as land zoning, a new shopping center, and the allocation of municipal resources. For community members, these divisions of topics into local and national issues were perhaps self-evident. From a sociological perspective, it remains to be understood why and how some issues came to be understood as locally relevant, whereas others did not.

The terms *nonresponse* and *apathy* are used in sociological literature to refer to social movement nonparticipation. The term *nonresponse* seems more neutral, but the label *apathy* carries a connotation of deficient moral qualities such as laziness or greed that prevent people from engagement. If we look to the root of the term *apathy*, however, we might apply it differently.

Buddhist environmental philosophers Joanna Macy and Molly Young Brown write that "to be conscious in the world today is to be aware of vast suffering and unprecedented peril." Because of the degree of destruction of life and potential for destruction of life, "pain is the price of consciousness in a threatened and suffering world," and "apathy is the mask of suffering" (1998, 26, 27, 191). Macy and Brown then describe apathy in the United States with respect to environmental problems not as lack of caring, but as being more like its Greek root, *apatheia*, which means the refusal or inability to experience pain. Thus, apathy is closely allied to denial at its root. Philosopher and practicing psychotherapist Shierry Nicholsen describes apathy as "a way of adapting, of defending oneself in a situation that is utterly overwhelming and where there is no end in sight." Furthermore, she adds that "we all need this kind of protection in our current environmental situation, certainly, in which there is so much destruction of so many kinds, in so many places, affecting so many people and so many other creatures, and with no end in sight. No wonder environmental activists complain about the widespread apathy that meets their efforts to arouse concern" (2002, 147).

Social movement theory has usually assumed apathy or nonaction to be the norm and focused attention on why people participate in movements rather than on why they do not (Klandermas 1997). In contrast, Eliasoph examines the production of apathy in everyday conversation. Unlike the dominant social movement perspective that nonparticipation is the norm, Eliasoph asserts that apathy takes work to produce: "Simple apathy never explained the political silence I heard. Instead of 'apathy' was a whole underwater world of denials, omissions, evasions, things forgotten, skirted, avoided and suppressed—a world as varied and colorful as a tropical sea bed" (1998, 255). People do care; they just aren't able to express their caring in all social contexts. And as Cohen reminds us in his work on denial, "Passivity and silence may *look* the same as obliviousness, apathy and indifference, but may not be the same at all. We can feel and care intensely, yet remain silent" (2001, 9). This observation is supported by studies of empathy and prosocial behavior that find that both people who experience very low levels of empathy and people who experience very high levels of empathy are less willing to help victims in a variety of situations (e.g., Stotland et al. 1978). People with high levels of empathy for victims found encounters so painful that they avoided involvement rather than assisting someone in need.

In fact, the association of apathy with greed or lack of concern reflects the widespread social (and often sociological) assumption regarding why people are not responding, which I seek to problematize. As illustrated by the examples in this chapter, I use the terms *apathy* and *nonresponse* to describe the absence of the issue of global warming in conversations or politics. I use the term *apathy* synonymously with *nonaction* and *nonresponse*. Rather than seeing nonresponse as an issue of lack of information or concern, we can approach it as being a result of socially organized *denial*—that is, the active (albeit rarely conscious) organization of information about global warming in such a way that it remains outside the sphere of everyday reality, a "distant" problem rather than a local political issue. Cohen reminds us in his work on denial that "the most familiar usage of the term 'denial' refers to the maintenance of social worlds in which an undesirable situation (event, condition, phenomenon) is unrecognized, ignored or made to seem normal." Cohen highlights the importance of ambiguity. Denial operates in the slippage zone between knowing and not knowing: "we are vaguely aware of choosing not to look at the facts, but not quite conscious of just what it is that we are evading. We know, but at the same time we don't know." He contends that "denial is always partial; some information is always registered. This paradox or doubleness—knowing and not knowing—is the heart of the concept" (2001, 51, 5, 22).

Denial, this condition of "knowing and not knowing," works precisely *because* it becomes natural, like everyday life, and thus invisible. The very real fact that Bygdabyingar (like everybody else) lead busy lives and find themselves pulled in multiple directions by the needs of family, work, home, community, and rest, works to support the necessary ambiguity. In 2000–2001, the fact that there were no (or very few) particular identifiable moments in which one ought to have taken action served well to cover up and explain the fact that individuals weren't doing anything about the issue of global warming.[3] Here, as Eirik struggled with how to understand why others in the community were not doing more on the issue of climate change, he referred to the fact that people do lead busy lives and "have to choose what to be engaged with":

Kari I asked you earlier if you felt that you had a sense of personal responsibility for climate change, and you said that you did. I wonder if you have a sense of whether other people have a personal relationship to this?

Eirik I am not really so sure. Because it can be, we hear a fair amount about it in the newspapers and so forth, it is a topic of discussion—such that people have the sense that they are collectively responsible. It is possible, but then at the same time I think that there are many who, we don't care, we'll close our eyes to it. I think that there are many who do this.

Kari And why do you think that is?

Eirik They have to choose what to become engaged with, and these unpleasant things they decide not to become involved with, I can imagine.

As we continued our conversation about why more people were not involved, it is the fact that he knew that people *are* busy that made Eirik unsure of exactly what was going on:

Eirik I don't know. There are certainly many reasons. I can't really answer it, but I can imagine many people are a bit active in different things. It can be their work or their family or caring for their children or their grandchildren that they really prioritize at that time or that they have a hobby, some people work on their cars, others cut wood in the forest, and some are active with sports. We often have lots of things that we are involved with.

Kari That one has to split the time between.

Eirik Yes, that you have to divide your time between. And if you have family, then you have to take into account what your children want to do and what your partner wants to do, and then there are the responsibilities that you have.

Kari And you have to sleep a bit and eat . . .

Eirik (Laughs.) Yes, yes, yes . . . so it can be that we have already filled up our time with lots of things, that can also be an important reason. Many people spend a lot of time with their families. Watching the kids, taking them on a hike.

Like the term *apathy*, the notion of being "in denial" has a negative connotation—being incapable of comprehending something or behaving out of stupidity or ineptitude. I wish to clarify that a key point in labeling the phenomenon of no direct activity in response to climate change as *denial* is to highlight the fact that nonresponse is not a question of greed, inhumanity, or lack of intelligence. Indeed, if we see information on climate change as being *too disturbing* to be fully absorbed or integrated into daily life—as is explored with interview data in chapter 3—this interpretation is the very opposite of the view that nonresponse stems from inhumanity or greed. Instead, denial can—and I believe should—be understood as testament to our human capacity for empathy, compassion, and an underlying sense of moral imperative to respond, even as we fail to do so.

And thus, it is when we take this latter view of denial that we see how apathy is the mask of suffering. The perspective of denial draws attention to an increasingly relevant psychological predicament for privileged people in this globalized "information age" when capitalism creates both a wider divide between the material conditions of the lives of haves and have-nots and reorganizes space and time in ways that bring privileged people ever closer to the worlds of those people they exploit through cheap airfares and quality digital Internet images. As Ulrich Beck and Anthony Giddens tell us, these new conditions of modern life have profound implications for our understanding of self and place, and for social action. For the well-educated and "nice" people who live in relative security, these circumstances make for disturbing contradictions. Under the surface of the rosy, serene picture of their lives are dark concerns. We move next to hear their voices.

3
"People Want to Protect Themselves a Little Bit": The Why of Denial

> There is concern that public ignorance and illiteracy about global environmental issues is leading to misinformed views, apathy, ill-considered calls for government action, and little change in personal behavior. This view of the relations between public knowledge, values and actions accords with what has been described as an information deficit model: Ignorance about climate change is preventing appropriate public action.
> —Harriet Bulkeley, *Common Knowledge?*

I am far from the first person to be puzzled by public silence in the face of climate change. On the contrary, environmental sociologists (e.g., Ungar 1992; Kempton, Boster, and Hartley 1996; Dunlap 1998; Rosa 2001; Brechin 2003, 2008), social psychologists (Halford and Sheehan 1991; Stoll-Kleeman, O'Riordan, and Jaeger 2001; Kollmuss and Agyeman 2002; Lorenzoni, Nicholson-Cole, and Whitmarsh 2007; Frantz and Mayer 2009), and public-opinion researchers (Saad 2002; Brewer 2005; Nisbet and Myers 2007) have alike for some time identified such public "apathy" as a significant concern. Possible explanations abound. I begin this chapter by examining existing explanations for why people have failed to respond to climate change from environmental sociology, psychology, and the field-of-risk perception points of view. I use the voices of community members to talk back to these dominant explanations for public silence. Although lack of information and lack of concern have been described as reasons why people do not respond to global warming, reasons for silence also come from people who are both informed and concerned about it. Here I explicitly shift from an information deficit model to a focus on the importance of emotion, social context, political economy, and social interaction in shaping how people relate to global warming.

"If People Only Knew"

For nearly twenty years, the majority of research on climate change from these disciplines presumed information was the limiting factor in public nonresponse. The thinking was that "if people only knew the facts," they would act differently. These studies emphasized either the complexity of climate science or political economic corruption as reasons people do not adequately understand what is at stake. Not surprisingly, given the extensive survey data on the public's lack of knowledge regarding climate change, the dominant theme of research from fields as widespread as science and risk communication, environmental sociology, and psychology has been the public's lack of information and knowledge as a barrier to social action. Systematic reviews of surveys and polling data by Thomas Brewer (2005) and Matthew Nisbet and Teresa Myers (2007) describe widespread misunderstanding regarding climate science extending back into the 1980s. Ann Bostrom and her coauthors write that "to a significant degree the effectiveness with which society responds to this possibility [of climate change] depends on how well it is understood by individual citizens. As voters, citizens must decide which policies and politicians to support. As consumers, they must decide whether and how to consider environmental effects when making choices such as whether our resources are most efficiently deployed by using paper or polystyrene foam cups" (1994, 959). John Sterman and Linda Sweeney (2007) similarly point to the complexity of atmospheric models as a limitation for both public understanding of climate change (even among highly educated people) and policy development. Noting "widespread misunderstanding" of how climate models work, the authors then link this conceptual failure to the lack of climate policy.

This assumption that "if people only knew," they would act differently—that is, drive less, use less electricity, or "rise up" and put pressure on the government—is widespread in popular discourse and environmental literature and underlies work from psychology, social psychology, and sociology. Psychologists and social psychologists have described flawed cognitive and mental models that limit people's ability to grasp what is going on, and sociologists have documented the manipulation of climate science (especially in the United States) and the media's role in misinforming the public by magnifying the perception of uncertainty. Sociologists have also conducted opinion polls highlighting the lack of public understanding of climate science and espousing the need for greater awareness.

The "conceptual challenges" surrounding global warming have been understood primarily in terms of the limitations of individual psychology (i.e., mental models, confirmation bias) or of media framing (see e.g., Ungar 1992; Bell 1994; Boykoff and Boykoff 2004; Armitage 2005; Dunwoody 2007; Boykoff 2008a, 2008b). Psychologists have described the power of "faulty" decision-making powers such as "confirmation bias" (Halford and Sheehan 1991). Bostrom and her coauthors describe how effective public response is limited because "lay mental models of global climate change suffer from several basic misconceptions" (1994, 968). Psychologists Grame Halford and Peter Sheehan write that "with better mental models and more appropriate analogies for global change issues, it is likely that more people, including more opinion leaders, will make the decision to implement some positive coping action of a precautionary nature" (1991, 606). From another angle, social psychologists consider "affect" to be the positive or negative evaluation of an object, idea, or image. Similar to emotions but not as "full blown," affect has been shown to powerfully influence both information processing and decision making. Work in the area of risk perception and affect in the United States and Great Britain by Irene Lorenzoni and her colleagues found that "the terms 'global warming' and 'climate change,' and their associated images, evoked negative affective responses from most respondents. Personally relevant impacts, causes, and solutions to climate change were rarely mentioned, indicating that climate change is psychologically distant for most individuals in both nations" (Lorenzoni, Leiserowitz, De Franca, et al. 2006, 266). Other work describes how confusion results from the fact that people relate to global warming through other existing generalized frames, what researchers call "mental models," and thus see it as an "ecological problem" in general, "air pollution" or "ozone depletion" (Stern, Dietz, and Guagnano 1995; Dunlap 1998).

Researchers have lamented the confusion between global warming and the ozone hole (e.g., Bell 1994; Bostrom, Morgan, Fischoff, et al. 1994; Read, Bostrom, Morgan, et al. 1994), investigated the role of media framing (Ungar 1992; Bell 1994; Brossard, Shanahan, and McComas 2004; Carvalho 2005), and described how understanding global warming requires a complex grasp of scientific knowledge in many fields. Harriet Bulkeley describes how in the dominant view people are

presented as individual agents acting "rationally" in response to information made available to them. According to the information deficit model of public response to environmental issues, the public needs to be given more knowledge about environmental issues in order to take action. . . . In this approach the contextual dimensions of environmental concern are ignored so that public perceptions are seen as stable, coherent, and consistent and to exist within individuals rather than being located within the inter-subjective contexts of institutions and discourse. (2000, 315–316)

A second body of scholarship points to relationships between political economy and public perception. Here scholars have identified the fossil fuel industry's influence on government policy (the United States provides prominent examples), the tactics of campaigns by climate change skeptics (McCright and Dunlap 2000, 2003; Jacques 2006; Jacques, Dunlap, and Freeman 2008; Jacques 2009), how corporate control of media limits and molds available information about global warming (Dispensa and Brulle 2003), and even the "normal" distortion of climate science through the "balance as bias phenomenon" in journalism (Boykoff 2008a, 2008b). Such political economic barriers presumably have far-reaching and interactive effects with the other factors discussed. Yet note that explanations for public nonresponse that highlight corporate media and campaigns by climate change skeptics also implicitly direct our attention to a lack of information as the biggest barrier to engagement, though for different reasons.

It is possible that the people of Bygdaby and their counterparts in the United States and around the world have paid little attention to the issue of climate change because they are too poorly informed to realize the potential danger or to be able to make connections between their daily activities and global warming. In my interviews and observations, there is certainly some evidence of general confusion about the basic facts. For example, in my time at Bygdaby residents in several instances confused global warming with the ozone hole. In one discussion about climate change, Sigurd, a member of the Labor Party, referred to climate change as coming from "holes in the atmosphere" (although he earlier correctly described climate change as coming from carbon dioxide emissions). There were other instances in which residents did not demonstrate a clear understanding of the process. One afternoon I joined a regular walking group of older folks on their tour of town. The unusual weather came up as a topic. I asked Maghild, a woman in her late sixties, if they went walking year round. Maghild replied, "Yes, so long as it isn't too slippery. You know this year is very unusual. I have never experienced it before. It can be 20 below and a meter of snow this time of year. Last year wasn't so warm. This is a very unusual year. And I was talking to a man who is much older than I am, and he doesn't remember anything like it either. They say it is because of climate change, the emissions of . . ."—here she let her sentence trail off, not sure what was being emitted that caused climate change. A half-hour later the topic came up again, and again she wasn't sure exactly what was being emitted to cause climate change: "What a good day for a walk," someone in the group commented. Maghild replied, "Yes, but very unusual weather, they talk of climate change, that the release of . . ."

The public may lack information, but is this fact limiting greater public interest, concern, and political participation? Despite cases in which people in Bygdaby lacked information, this absence of information didn't seem to be the limiting factor in their reaction to climate change. Indeed, in both the conversations I just described, the individuals *were* concerned about global warming, despite their confusion or missing information. As Daniel Read and his coauthors (1994) point out, only two simple facts are critical to understanding climate change. First, if significant global warming is occurring, it is primarily the result of an increase in the concentration of carbon dioxide in the earth's atmosphere. Second, the single most important source of carbon dioxide addition to the earth's atmosphere is the combustion of fossil fuels, most notably coal and oil. Norwegians are among the most highly educated people in the world, and, "availability heuristics" and "mental models" aside, the basic fact that burning fossil fuels releases carbon dioxide and contributes to global warming is hardly a technical piece of information. If people don't know, then why not?

A key problem with information deficit models is that they do not account for the behavior of the significant number of people who *do* know about global warming, believe it is happening, and express concern about it (Hellevik and Høie. 1999), as appears to be the case for the majority of Bygdaby residents as well as for a sizeable percentage of the U.S. population. Recent Gallup data on the United States indicates that now some 80 percent of Americans report that they do understand global warming (Newport 2008). Yet as we saw in the discussion of the background noise phenomenon in chapter 2, this increased understanding has mysteriously failed to translate into either greater concern or concrete action.

Another approach applies psychological theories on cognitive dissonance, efficacy, and helping behavior to climate change (see, e.g., Stoll-Kleeman, O'Riordan, and Jaeger 2001; Kollmuss and Agyeman 2002; Lorenzoni, Nicholson-Cole, and Whitmarsh 2007). Leon Festinger's (1957) concept of cognitive dissonance describes "dissonance" as a condition that emerges when an actor has two thoughts (cognitions) that are inconsistent. This dissonance is an unpleasant condition that people seek to resolve, often through changing one of their cognitions. Studies drawing on these frameworks point to multiple factors that would seem to "complicate" how people process information on climate change. This complication explains Paul Kellstedt, Sammy Zahran, and Arnold Vedlitz's (2008) finding that increased levels of information about global warming have a negative effect on concern and sense of personal responsibility. They noted in particular how respondents who are better informed about climate change express less rather than more responsibility for the problem. And they also found that "in sharp contrast with the knowledge-deficit hypothesis, respondents with higher levels of information about global warming show less concern" (2008, 120). These findings are in accordance with cognitive dissonance because people with low self-efficacy will be likely to deny responsibility and concern because unless they feel able to do something about the problem, an awareness of concern or sense of responsibility would be a conflicting cognition. Jon Krosnick and his colleagues (2006) similarly observe that people stop paying attention to global climate change when they realize that there is no easy solution for it. Instead, many people judge as serious only those problems for which they think action can be taken. In another highly relevant application, Cynthia Frantz and Stephan Mayer (2009) apply a classic model of helping behavior to the public response to climate change. Based on the criteria of this model, the authors note that climate change is difficult to notice and is marked by a diffusion of responsibility and that there are psychological costs of acting, each of which inhibits the likelihood of individual response.

Widespread public belief that climate change is happening clearly contradicts the assumption that lack of information is the key variable behind public apathy. Although there presumably exist at least some climate skeptics in Bygdaby, this widespread belief is congruent with the findings of national-level Norwegian studies of public response to climate change, which indicate that a majority of Norwegian citizens are concerned about climate change (Hellevik and Høie 1999). In Norway, there are far fewer people who do not believe that climate change is happening than in places such as the United States, where coal and oil industry–related organizations have waged large counter-campaigns (McCright and Dunlap 2000, 2003) and where President Bush himself openly questioned the validity of scientific data. Yet even in the United States, most likely the country with the highest percentage of climate change skeptics in the world, skeptics only make up a minority of the population.

In fact, as indicated earlier, in that winter in Bygdaby in 2000–2001 the sense that the weather was very different from earlier times was considered "common knowledge" in the community, and comments on the unusual weather were consistently linked with the possibility of climate change. People spoke often of the weather being "less stable" than in the past. Eirik, a community member in his early fifties who worked for the county, voiced a sentiment that was frequently heard:

Eirik And it has been quite clear since the end of the 1980s, early '90s. There is a totally different climate here now than when I was a child.
Kari Really?
Eirik Oh, yes. Much colder winters and more stable [in the past]. Even though there have always been small changes, it is clear that there are now significant differences. And at the same time I see a connection with all the things that we hear from Africa and other continents about climate changes, famine, dry spells, I feel that we learned this in school, that these climate gases, they are at a certain level, and we can measure that they are so much higher than they have been.

Although I did meet one person who said he was not concerned about global warming and a few who raised the possibility of doubt, I did not meet anyone in Bygdaby who dismissed it as an insignificant issue. Hilde, a member of the Farm Women's Association in her sixties, described her reactions to global warning: "We think it's a bit odd, you know. The way I remember winters, or winters before, you know there was always lots of snow, and it was cold the entire winter, you know." Lars, another local political leader, who stated that he believed climate change was happening, expressed some reservations about holding human beings responsible for it. Nonetheless, he said that caution was the wisest approach:

It's like politics. You have to choose who you trust. And I surely believe that there is climate change because we are constantly having new records, so that can't be explained away. But whether it is pollution that is responsible or whether it is happening on its own, that's too difficult to know. I don't know. There are scientists who say that it is coming no matter what. But of course we shouldn't take that chance. We shouldn't pollute more than necessary here in this world.

Finally, the notion that well-educated, wealthy people in the Northern Hemisphere do not respond to climate change because they are poorly informed not only appears to be inadequate to explain the nonaction in

Bygdaby and much of the United States, but also fails to capture how in the present global context "knowing" or "not knowing" is itself a political act. All nations emit carbon dioxide and other climate gases into the common atmosphere, though the wealthiest 20 percent of the world's population is responsible for more than 80 percent of cumulative global greenhouse gas emissions. Nevertheless, global warming will precipitate the most extensive and violent impacts against the poor and people of color of the globe. Poor nations in Asia, Africa, and Latin America already experience more than 90 percent of the world's disasters and disaster-related deaths. Thus, not only is global climate change the most serious environmental problem of our time, but it is also a highly significant human rights or "environmental justice" issue (Agarwal and Narain 1991; Baer, Harte, Haya, et al. 2000; Roberts 2001; Athanasiou and Baer 2002; Donohoe 2003; Pettit 2004; Roberts and Parks 2007). Industrialized nations of the Northern Hemisphere emit greenhouse gases disproportionately to the global airshed, but a lack of resources and infrastructure place poor nations most at risk (Watson, Zinowera, and Moss 1998). It is highly significant that Norwegian wealth comes directly from the production of oil and that its economy flourishes with the current level of carbon dioxide emissions. I noted earlier that Norway is the largest oil producer in Europe and (as of 2009) the world's fifth-largest oil exporter (United States Energy Information Administration 2009). More than one-third of Norwegian national revenue is generated from the petroleum industry. Expansion of oil production in the 1990s enhanced the already high standard of living in Norway. These developments occurred during the same time that the Norwegian government backed away from Kyoto targets and that the percentage of the public that was "very much worried" about global warming dropped from 40 to 10 percent.

Given that Norwegian economic prosperity and way of life are intimately tied to the production of oil, ignoring or downplaying the issue of climate change serves to maintain Norwegian global economic interests and to perpetuate global environmental injustice. It is easy to see power operating when key political and economic decision makers negotiate contracts with Shell, British Petroleum, and Exxon and when representatives of nation-states negotiate emissions-trading strategies. Yet the people I spoke with in Bygdaby played a critical role in legitimizing the status quo by not talking about global warming even in the face of late winter snow and a lake that never froze. The absence of these conversations worked to hold "normal" reality in place.

Former Norwegian minister of the environment Børge Brende has expressed that "Norway is one of the countries in the world that has benefitted most from fossil fuels. This gives us a special responsibility in the politics of climate change, especially with respect to poor countries" (Hovden and Lindseth 2002, 143). Despite its reputation for environmental leadership, Norway has tripled its production of oil and gas in the past ten years. Under the Kyoto Protocol, Norway promised to limit greenhouse gas emissions to a maximum of one percent above 1990 levels. Instead, at the time of my stay in 2001, total Norwegian carbon dioxide emissions were 42.4 million metric tons—an increase of 7.2 million tons or 20 percent from the 1990 level of 35.2 million tons (Statistisk sentralbyrå 2002). Norwegian researchers Eivind Hovden and Gard Lindseth note that "Norway, an already wealthy and highly developed country, built a very significant fortune in the 1990s from the very activity that has made stabilization of CO_2 emissions next to impossible" (2002, 163). By 2008, the emissions figure had climbed to 53.8 million tons. This critique is echoed by Norwegian climate policy analyst William Lafferty and colleagues. In their review of progress on sustainability, the authors note that despite Norway's early international leadership on the issue under Gro Harlem Brundtland, as of 2006 "the Norwegian Sustainable Development profile is long on promise and short on delivery" (2007, 177). Lafferty and colleagues point directly to the role of oil wealth in the shifts in national policy: "In our view, a major reason for this 'reluctance' is the increasingly dominant role of the petroleum sector in the Norwegian economy. The impact of the petroleum economy on the will to pursue sustainable production and consumption in Norway has been massive. The prospect of steadily increasing state revenues from petroleum and gas activities has directly 'fueled' the politics of both 'business as usual' and increasing welfare benefits" (2007, 186). As of 2008, the oil and gas industry accounted for 26.6 percent of the Norwegian carbon dioxide emissions.[1]

The notion that people are not acting against global warming because they do not know about it reinforces a sense of their innocence in the face of these activities, thereby maintaining the invisibility of the power relations that are upheld by so-called apathy regarding global warming. Within this context, to "not know" too much about climate change maintains the sense that if one *did* know, one would act more responsibly. This can be seen as a classic example of what Susan Opotow and Leah Weiss call "denial of self-involvement": "Denial of self-involvement minimizes the extent to which an environmental dispute is relevant to one's self or one's group. . . . By casting themselves as 'clean' and insignificant contributors to pollution, they assert their non-relevance to environmental controversy" (2000, 485). Stanley Cohen similarly observes that

> The psychology of "turning a blind eye" or "looking the other way" is a tricky matter. These phrases imply that we have access to reality, but choose to ignore it because it is convenient to do so. This might be a simple fraud: the information is available and registered, but leads to a conclusion which is knowingly evaded. "Knowing," though can be far more ambiguous. We are vaguely aware of choosing not to look at the facts, but not quite conscious of just what it is we are evading. We know, but at the same time we don't know. (2001, 5)

Citizens of wealthy nations who fail to respond to the issue of climate change benefit from their denial in economic terms. They also benefit by avoiding the emotional and psychological entanglement and identity conflicts that may arise from knowing that one is doing "the wrong thing," as I discuss more fully later.

Most work on concern, knowledge, and perception has taken the form of large-scale surveys. Data from interviews and ethnographic observation can yield information on meanings and relationships between thinking and feeling in everyday life. Indeed, these studies emphasize the complexity of people's response to climate change—how people seem to know and not know about it at the same time—and point to the emotional ambivalence that characterizes denial. As Bulkeley notes, "Confusion, doubt and a degree of illiteracy concerning climate science did not prevent focus group participants from locating this global issue in their backyard," and "there is a need to move from a narrow conception of public knowledge towards recognition of the complex, fluid and contradictory nature of public understanding of global environmental issues" (2000, 329).

It is important to note that although I offer significant criticisms of the information deficit model, the idea that people do not need any information regarding climate science to develop concern or engage in action is not one of them. What is important to recognize, however, is that information alone is not enough to produce action. "Information," like caring (see the next section), cannot be thought of in generic and isolated blocks of "facts" with universal meaning and significance across all communities. Instead, information is socially structured, is given social meanings, and must be understood in social context. As I explore in chapter 4, information on climate change may be accepted, resisted, navigated, and interpreted differently depending on the sense of efficacy, self-esteem, and social support of the individuals receiving it.

"If People Only Cared"

Just as we can ponder whether people in Bygdaby really knew the facts in 2000–2001, it is also possible that people in this small town paid little attention to climate change because they simply didn't care about it. This explanation for inaction describes people as too selfish, too individualistic, or too greedy to think about the well-being of others, whether the others are future generations, citizens of other parts of the world, or nonhuman organisms and ecosystems. As awareness of climate change increases worldwide and information deficit explanations have become less prominent, more social scientists have more recently turned to explanations for inaction that hinge on inadequate caring.

Given so little visible social action regarding climate change in Bygdaby, not to mention elsewhere, such as the United States, one might easily come to such a conclusion. Yet all but two of the people I spoke with in my year living in Bygdaby voiced significant concern about climate change. During a portion of our conversation in which we discussed the production of climate gases and climate change, Eirik expressed a sense of concern about future impacts: "I see that we do lots of things that most certainly cannot continue. It will work for a while, but sooner or later it isn't going to work. So I am worried in any case for that which will happen."

In my conversation with Øystein, I asked whether, when he thought of environmental issues, there was one that stood out for him in particular. He indicated that the stand-out issue was climate change, and he was clearly concerned about it: "Most I would say emissions. Emissions in the atmosphere. Perhaps that is where it is dangerous, where we are doing poorly today. You know, emissions are going right to hell. It is dramatic."

In order to avoid the possibility that people were expressing concern about climate change simply for reasons of social acceptability, I often asked interviewees whether they thought *other people* were concerned about global warming. Arne, a man in his early sixties, reflected simultaneously on both his own fears and my question of what he thought other people were thinking: "I don't completely know what I shall think of it. But regardless, I believe that many believe that it's wrong that we are changing nature so much. I have the sense that most people believe that nature knows best. I believe that. So, basically, I think that people are worried about climate change. They don't know rightly what to say." These are hardly the thoughts of a person who callously disregards the future or the lives of others. A lack of caring per se did not seem to be a problem in Bygdaby.

Furthermore, public-opinion data on concern are hardly linear. As evidence for climate change pours in and scientific consensus increases, interest in the issue throughout many Western nations paradoxically declined during the 1990s and into the early 2000s (e.g., Immerwahr 1999; Hellevik 2002; Saad 2002). For example, Gallup polls for the United States show that the percentage of people who "personally worry a great deal about global warming" dropped from 35 percent in 1989 to 28 percent in 2001, and the percentage who worry "not at all" rose from 12 to 17 percent during the same time period (Saad 2002). Biannual national-level survey research in Norway even more dramatically finds a significant and steady downward trend in public interest and concern about global warming, with the percentage of respondents who replied that they were "very much worried" declining steadily from 40 percent in 1989 to less than 10 percent in 2001 (Hellevik 2002; Barstad and Hellevik 2004).

If we consider the voices of community members alongside patterns of survey data, the notion that people do not respond because they do not care about climate change appears at best to be an incomplete explanation. Such results are deeply troubling to our enlightenment sensibilities in which we presume that knowledge will lead to rational action. Indeed, the basic premise of an enlightened, democratic, and modern society is that information (especially scientific information) will lead to concern and response on the part of the public and public institutions. Yet the case of climate change poses a challenge to this paradigm (Norgaard 2006a, 2006b, 2009, 2011; Jacques 2006; Jacques, Dunlap, and Freeman 2008; Jacques 2009). Instead, relationships among caring, knowledge, and action point us to another set of questions about whether and under what circumstances information leads to concern or action (Krosnick, Holbrook, Lowe, et al. 2006; Kellstedt, Zahran, and Vedlitz 2008).

Hierarchy of Needs

A related explanation for public nonattention to climate change emphasizes a kind of "hierarchy of needs" (Maslow 1970) in which people focus on immediate needs first and long-term needs later. In this line of reasoning, people cannot think about climate change because they are too consumed with solving the problems of the present. Although there is no doubt that the hierarchy-of-needs approach holds weight—indeed, each of us is clearly confronted with more issues than our attention can handle—this approach alone is also insufficient to explain public apathy on the larger social level. Individuals in a particular social context may express the feeling that they don't have time or may have a limited ability to respond or both. Yet from a sociological standpoint, this information tells us about that particular society's social norms and limits of concern. "Needs," however real they may feel, are, in affluent places such as Bygdaby where nobody lives "on the edge," a reflection of social facts and local social and cultural norms. In his work on cognitive sociology and the social organization of concern and caring, Eviatar Zerubavel writes, "After all, only through being socialized does one come to know whether the concern about feeding one's dog should come before or only after the concern about feeding the homeless, or whether one ought to be more concerned about the well-being of fellow American businessmen in Southeast Asia or the Southeast Asian refugees living in one's own neighborhood" (1997, 47).

People in Bygdaby may feel that they can't use less fossil fuel because they "need" to be able to drive their kids to soccer practice or to take an annual trip to Greece, but these kinds of needs are very much a product of social context. In the words of Eirik, who had lived with his wife in Africa as a missionary for several years,

We shouldn't consume so many resources, drive so much, or travel so much by air. We know that it is bad because it increases carbon dioxide levels and creates a worse situation. But at the same time, of course, we want to go on vacation; we want to go to the South; we want to, well, live a normal life for today. So many times I have a guilty conscience because I know that I should do something or do it less. But at the same time there is the social pressure. And I want for my children and for my wife to be able to experience the same positive things that are normal in their community of friends and in this society.

Here Eirik struggles with his own sense of right and wrong behavior and the social pressure to need a life that is more "normal" for his society. These pressures are significant.

Another facet of the hierarchy-of-needs explanation is that the issue of climate change is too abstract, one that affects people only in the future and thus is farther down on the hierarchy of needs. Yet warm temperatures and the absence of snow were hardly abstract issues of the distant future that winter of 2000–2001 in Norway. Numerous noticeable effects of unusual weather were commonly interpreted as possible impacts of climate change. As previously mentioned, skiing is an activity with a great degree of cultural importance in Norway, and when snow came late that particular winter, causing a two-month delay in the opening of the ski area, there were significant, tangible economic and cultural effects for the community.

More important, like "needs," perceptions of what is near and far, relevant and abstract are themselves socially structured (Zerubavel 1997). Norms of attention, in Bygdaby as elsewhere, organize perceptions of reality (see chapters 1 and 4). Indeed, C. W. Mills's (1959) work on the sociological imagination, a concept that is at the very heart of sociology, is fundamentally about whether individuals "see" connections between their lives and politically relevant events in the world around them.

In her work on apathy in the United States, sociologist Nina Eliasoph notes that for many of her respondents, battleships in their front yards and toxic wastes in their neighborhood were not considered sufficiently "close to home" to warrant action, yet whales in the North Pole were (1998, 2). If climate change is felt to be an "abstract" issue in the community, this fact reflects a disjuncture between the local sense of time and place described in chapter 1 and the sense of time and place that would be needed to conceptualize climate change for it to seem "real." Norwegian sociologist Ann Nilsen interviewed young people in Bergen about environmental problems and their sense of the future. Nilsen similarly concludes that

> the most serious consequences from damaging the environment, are long term. In societies such as the contemporary Western world where thinking and attention span are aimed at the extended present, or the immediate future, environmental problems of the magnitude that climate change represents, for instance, will be difficult to find solutions to, also because of a general time horizon involving less attention to the long-term future. (1999, 176)

A community's sense of the past, present, and future are not just "there," like a political imagination; they are collectively constructed. In fact, there is virtually no evidence to support the perspective that climate change just does not pan out in a hierarchy of needs. For example, the European nation that is threatened most by sea-level rise, the Netherlands, ranks at the very bottom of level of concern regarding climate change in ACNielsen's 2007 global study of nations (ACNielsen 2007).

And Sammy Zahran and coauthors find that in the United States "respondents living within 1 mile of the nearest coastline at negative relative elevation to the coast are less (not more) likely to support government-led climate initiatives" (2006, 783). Again in relation to the hierarchy-of-needs argument, consider the negative relationships between wealth and concern exhibited in individual, state, and national data (e.g., O'Connor, Bord, Yarnal, et al. 2002; Zahran et al. 2006; Sandvik 2008). Consider the findings of Hanno Sandvik (2008), who examined a cross-national sample of data on public concern for climate change from 46 countries. Sandvik hypothesizes that public awareness and concern regarding climate change is not a function of scientific information alone, but of psychological and sociological factors as well. He observes a negative association between concern, on the one hand, and national wealth and carbon dioxide emissions, on the other, and notes a "marginally significant" tendency that nations' per capita carbon dioxide emissions are negatively correlated to public concern. Sandvik writes, "These findings suggest that the willingness of a nation to contribute to reductions in greenhouse gas emissions decreases with its share of these emissions." He concludes that such a relationship is "in accordance with psychological findings, but poses a problem for political decision-makers" (2008, 333). Although Sandvik is the first to test explicitly a relationship between wealth and concern cross-nationally, his findings are in accordance with earlier work across spatial scales from the individual to the nation-state. For example, Zahran and colleagues (2006) found that citizens residing in U.S. states with higher emissions of climate gases are somewhat less likely to support climate change policies. Robert O'Connor and his coauthors (2002) found that higher income negatively affected participants' willingness to take some voluntary actions such as driving less. An inverse relationship between wealth and concern is also reported in Riley Dunlap's 1998 cross-national research, but with a smaller sample of nations. Although Dunlap does not give this relationship much discussion, he notes that "despite the lower levels of understanding among citizens of the poorer nations in our study (Portugal, Brazil and Mexico), residents of these nations typically express more concern over global warming than do those in the more affluent nations (Canada, U.S., Russia)" (1998, 488). Furthermore, there are no examples of the reverse relationship, in which higher income is positively correlated with concern for global warming or with support of climate-protection policy.

All Is Well

It is also possible that people in Bygdaby, like others around the world, pay little attention to climate change because they believe that the government will take care of things, that international agreements on emissions reductions will be reached, and that all will go well. This perspective might be a variation of either "faith in government" or "technological optimism." In past decades, the Norwegian government was visibly involved in the issue of climate change. Perhaps residents feel that things are in good hands. Norwegian sociologist Ottar Hellevik explores the possibility of faith in the government as a causal factor behind the national pattern of declining concern about climate change since 1980. He does not, however, interpret optimism as the force behind the trend in declining concern:

> Results from the Monitor surveys tend to contradict such a trend of optimism, however. The percentage disagreeing with the proposition—"When negative environmental conditions are revealed, business takes the problem seriously and cleans up as soon as possible"—rose from 45 percent in 1995 to 56, 59 and 55 percent in 1997, 1999 and 2001 respectively. The public thus seems to have become *more* rather than *less* skeptical with regard to the environmental status of business leaders. Further, when queried in 2001 as to anticipated development trends for selected areas of society, only 14 percent of the population believed that the situation with regard to environment and pollution would improve, while 31 percent expected deterioration (43 percent reply, "no major changes" and 10 percent "don't know"). (2002, 13–14)

Although I suspect it is possible that some community members in Bygdaby felt a sense that "all was well" at the beginning of this century, I found no one who endorsed this perspective. Instead, the expressions of concern given in earlier quotations indicate that a significant number of individuals felt that all was not well. My field notes offer another example on the local level from a conversation I heard one evening:

> It was a mid-November evening, and Sam and I joined our neighbors at a local dance organized by a group that Anne says has been around about 30 years. A band played folk dance music, and the atmosphere in the room was friendly and happy. Most people were pretty good dancers. Sam and I danced a reinlander, a waltz, and a few other dances. There was a pause in the music, and we went back to our table. Arne said to Torstein but clearly and facing me so that I could understand: "You know the second time that I got married, in the 1970s I didn't really want to. I didn't feel that I should have any more children, didn't know what kind of a world my grandchildren would have, so didn't feel that it

was right to have children." Then he said, "All these meetings in The Hague [referring to the recently failed climate talks], and what has come of it? Nothing." (field notes, November 18, 2000)

People may have *hoped* that all was well, but for these residents the jury was still out.

Political Alienation

Finally, a corresponding but opposite explanation for the "all is well" hypothesis is that perhaps people are *so* disempowered that they are not responding to *anything*. It is possible that the lack of response to climate change is not specific to the issues raised in this case, but part of a general phenomenon of public apathy about and lack of interest in the environment. In the United States, for example, voting rates and faith in democracy are lower than in Norway, so the hypothesis of political alienation may hold more weight there. Indeed, there is some evidence for political alienation in Norway as well. Arne, something of an environmental philosopher, describes his sense that people are disillusioned with the concept of progress: "I am afraid that there is less optimism today than in the past. There has been more optimism. There was more optimism before. We can look at it in terms of philosophy. We talk about modernity and the modern time or whatever it's called. People are disillusioned, they no longer believe in the great notion of progress. So Norway isn't the fantastic country like that anymore." Peter, a man active in the opposition to the European Union, said he felt Norwegians were not optimistic:

Peter Well, no, I don't have a lot to say about whether Norwegians are optimistic or pessimistic, but I think it's probably that people are more and more pessimistic; yeah, I think probably so.
Kari Why would that be?
Peter They see that it doesn't matter, that the development of society is driving over them.
Kari Hmm, I see. Have you experienced this?
Peter Yes, I have experienced it. I experience it, I experience it, and I experienced it in the last years when people said no [to the European Union]. In the last municipal election—the local election—I experienced it a lot. "No, there's no point. It doesn't matter."

Yet, as I discussed in chapter 2, there were in fact high levels of political activity in Bygdaby in 2000–2001, on issues ranging from local topics such as the development of the downtown, new roads, and a proposed shopping mall to national and international issues such as opposition to the European Union. Other social indicators of political trust such as voting behavior do not indicate that political alienation is so prevalent that it alone can explain the public silence on climate change. As mentioned in chapter 1, rates of voter participation on local and national levels were relatively high in Norway. Was there something about the issue of climate change that made engagement more difficult? If so, can an in-depth examination of how people respond to climate change tell us something more about the meaning and process of political alienation? Or can it perhaps especially bring insight into the dynamics of politics in the new terrain of risk societies?

Contrary to widespread assumptions that people fail to respond to global warming because they are too poorly informed, are too greedy or too individualistic, and suffer from incorrect mental models or faulty decision-making processes, the people I spoke with expressed feelings of deep concern and caring and a significant degree of ambivalence about the state of the world. Instead, as I listened, the residents of Bygdaby told me many things about why it was difficult to think about global warming. I mentioned earlier the words of one person who held his hands in front of his eyes as he spoke: "People want to protect themselves a bit." Knowing about global warming raised fears for the future, feelings of helplessness, and feelings of guilt, some of which were in turn threatening to individual identity. Yet emotions, despite their apparent salience in how people process information on climate change, are missing from the current scholarly discussion about nonresponse. If in Bygdaby emotions worked to prevent engagement, how exactly did this happen? By what mechanism did this process take place? Here I use community members' voices to lay out a series of unpleasant emotions linked to thinking about climate change (see table 3.1).

Table 3.1
Troubling Emotions Associated with Thinking about Climate Change

---

Fear of loss of ontological security
Helplessness
Guilt
Threat to individual and collective sense of identity

Risk, Modern Life, and Fears Regarding Ontological Security

Automobile and plane crashes, toxic chemical spills and explosions, nuclear accidents, food contamination, genetic manipulation, the spread of AIDS, global climate change, ozone depletion, species extinction and the persistence of nuclear weapons arsenals: the list goes on. Risks abound and people are increasingly aware that no one is entirely safe from the hazards of modern living. Risk reminds us of our dependency, interdependency and vulnerability. Catastrophic risk is an even stronger reminder.
—Carlo Jaeger, Ortwin Renn, Eugene Rosa, and Thomas Webler, *Risk, Uncertainty, and Rational Action*

One day in mid-December 2000, my husband and I, avid skiers ourselves and disappointed with the lack of snow in Bygdaby, decided to take the train a few hours away to a neighboring community and go skiing. The temperature was about −5°C, and the sun was shining brightly on the bare fields surrounding our house as we loaded our skis into the taxi and drove down the road to the train. "Do you like to ski?" I asked our driver. "Oh yes, but I don't do much of that anymore," he replied. He was in his late fifties, but this age doesn't imply much for a Norwegian because I have been outskied by many people older than seventy! "When I was a kid, we would have skis on from the first thing in the morning to the end of the day. I grew up in Mykdalen [a community about 20 kilometers from Bygdaby], and there was so much more snow back then. We had snow from October to May. You know the parade on May 17, that was always in the snow." I ask if the weather really has changed all that much. "Oh yes," he says, "The summers were warmer, and the winters were colder with more snow. In the summer, we would spend a lot of time swimming in the river. We have seen so many changes. The climate is changing quite a lot." "That's a bit scary," I say. "Yes," he agrees. "When you think of how much has changed in my fifty years, I was born right after the war, how much has changed in my fifty years of existence, it is very scary."

Large-scale environmental issues in general and global warming in particular threaten biological conditions, economic prospects, and social structure. The impacts of global warming on human society are predicted to be widespread and potentially catastrophic. At the deepest level, large-scale environmental problems such as climate change threaten individual and community senses of the continuity of life—in other words, they threaten what Anthony Giddens calls "ontological security." "Ontological security" refers to the confidence that most human beings have in the continuity of their self-identity and the constancy of the surrounding social and material environments of action" (1991, 92). Merely thinking about climate change raises a series of questions related to ontological security: What will Norwegian winters be like without snow? What will happen to farms in the community in the next generation? Will they, in the words of one Bygdaby resident, "begin growing oranges in Norway?" Robert Lifton writes of an increasing, "amorphous but greatly troubling sense that something has gone wrong with our relationship to nature, something that may undermine its capacity to sustain life" (1982, 21). If the climate continues to warm, how are people going to make a living and maintain their lifestyles in 50 years? In 100 years? Thinking about climate change was difficult for people in Bygdaby at the turn of the twenty-first century because it brought up fears connected to ontological security. In Arne's words, "I think it's a bit worrisome to lose one's roots or to lose connection with, yes, with the generations and with a place."

People have a need for meaning in their lives. The present environmental crisis threatens not only people's sense of how the world is (a "good place," as many want to believe), but also the meaning of their sense of the continuity of life, as Lifton discusses at length. Joanna Macy and Molly Young Brown describe how we look away, "lest we drain our lives of meaning" (1998, 28). As Øystein told me, "I think maybe that most people think so little about climate change because they are afraid we are on the wrong track. That it could go badly. But we don't know . . . that this is as far as it goes for Norway."

Øystein also expressed the concern that climate change is deeply threatening to one's sense of the continuity of life, commenting on the possibility that in as few as a hundred years it may not be possible to live on the planet:

We have now come so far that . . . you know, in one hundred years it's possible that the environment will be damaged to the point that it isn't possible to live on earth anymore, you know? You see, of course, that we have these holes in the atmosphere that climate change is coming from, you know? Now people are beginning to see that something is happening with nature because we haven't taken environmental questions seriously enough.

For Lise, mother of two and member of the Socialist Left Party, climate change was "one of the reasons I try to be active"—although she hadn't actually taken action related to climate change. "Yes, of course it is one of the reasons that I try to be active—try to influence, you know. I am very pessimistic when I think about these things."

Ulrich Beck describes the present phase of modernization as a "risk society," one that is characterized by a "growing embeddedness of risk in the public consciousness" (Beck, quoted in Jaeger et al. 2001, 13). Beck argues that modern society, through the use of scientific technologies, has created large-scale social and environmental risks that cannot be understood without scientific expertise. For him, the risk society is one in which ontological security is threatened in two different ways: first, modern life means pervasive exposure to risks that threaten our sense of the continuity and stability of our lives; and, second, in modern societies the social networks of human life have been eroded. For community members in Bygdaby in 2000–2001, both these threats to ontological security were visible concerns.

Feelings of Helplessness

Lise's comments about pessimism were connected to the often voiced sense of helplessness or powerlessness in relation to climate change. As she continued speaking, she described the many problems in the country, which she said could make her feel a bit "pffff" (meaning "yeah, whatever"):

Lise It's like that with the environment, with the women's movement, with the green [movement], and with war and peace and everything. There are lots of problems in the country. There's a lot that is negative; I feel a bit like, yeah, pffff! But when I have something that I am trying to do, something with, when you are trying to influence something, then it's like you can be optimistic anyway. But I think about this with the young people. Things can just explode around us all the way, and so it's good that I don't allow myself to think so far ahead.

Lise mentioned that one strategy for coping with these feelings is to get involved and that another is not to think too far into the future. As noted in chapter 2, another Bygdabyingar told me that it was "just a joke to get involved with that," meaning climate change. And others expressed a similar feeling:

Ingrid I think that there are lots of people who think, "I don't have that problem myself; I can't do anything about it anyway."
Kari They don't feel that they can do anything anyway?
Trudi I think that there are a lot of people who feel "No matter what I do, I can't do anything about that anyway."

Helplessness, then, was a second emotion that the topic of climate change evoked among Bygdabyingar in 2000–2001. Trying to think about this problem could be overwhelming. The problem seems so large: solving climate change will involve the cooperation and common work of people in so many different countries, governments are unable to reach agreement, and perhaps entire economic structures will have to change. Even if all of this change were to be achieved, all the carbon dioxide released up to the present will still continue to cause climate change. Thus, it is not surprising that rather than feeling that there is much that can be done, one resident, Maghild, a woman in her late sixties, pronounced that "we must take it as it comes." And Lene told me, "And of course it's climate change that is doing it. There isn't anything to be done about it." Beyond the dimension of powerlessness that comes from the situation itself but connected to that dimension is the possibility that those political and economic structures that have been set in place are inadequate to handle the problem. Thus, for some residents there was another layer to the feeling of powerlessness that comes from considering the possibility that neither one's government nor the world community at large can be relied on to solve this problem. Arne said that he was afraid that there was less optimism than before, and Peter felt that more people have the sense that "no, there's no point. It doesn't matter." This concern is connected to the growing general sense of helplessness throughout modern society worldwide. Carlo Jaeger and his colleagues note that "the adoption of 'risk' as the imprimatur of our age marks a significant refocusing of social thought. The foundation of Western thought since the Enlightenment—from Comte, Spencer, Marx, Parsons, Habermas and others—has been the expectation of progress, of continued improvement in the social world. The emergence of a 'Risk Society,' abruptly challenges that assumption" (2001, 15).

Beck writes that the "risk society means an epoch in which the dark sides of progress increasingly come to dominate the social debate" (Beck 1992, 2). As previously discussed, trust in the government has been relatively high in Norway, thus further highlighting the tension and significance of those instances where this trust is challenged.

Lise described the way in which her feelings of helplessness with respect to climate change were merging with her feelings regarding environmental risks in general. Although she confused climate change with the depletion of the ozone layer, she was clearly concerned. She vividly described a choice between the chemicals in sunscreen, which she feared may be carcinogenic, and what would happen if her son didn't get sufficient protection:

It is so icky, that I believe that the sun has become stronger. Yeah, so you go around with that inside you, we carry that all the time. But it was so strange, we went skiing two weeks ago and we needed to go into a store to buy sunscreen. So I went in and there were all these different factor levels. It gets so crazy, the whole thing. Because you know how the sun is, that it burns and makes the skin damaged, and can make you sick. I always put sunscreen on the kids, I think that we are probably predisposed to get burned. So I bought factor 21, which protects him, but factor 21, there are lots of chemicals in the cream, so then—you don't know—so you know how the whole things goes, right? So there was a man standing there watching me, what I was doing and trying to choose. "No, no, no," he said to me—I didn't know him at all—"you must take factor 8," he said, "because that one doesn't have so many chemicals in it, [and] it will protect you well enough." He was watching me because I was talking out loud in the store, you know, trying to decide which one to take. But it illustrates a bit of the apathy that we live with these days. And that mad cow disease, for example. Now I try to use less meat and more vegetables. And the next day you learn that no, vegetables, they have so many cancer-causing pesticides of this type and that type. So you are constantly reminded of everything that is dangerous and unhealthy and carcinogenic. So, you know, you are reminded of that all the time. That's when I think I have to just cut out and live as well as we can. Yeah. And find the middle path, the way through, all the time. I am preoccupied, absorbed, with it. It is exhausting!

After describing her attempts to choose the lesser of two evils, sunscreen or sunburn, Lise went on to mention other environmental issues that troubled her, from mad cow disease to pesticides on fruits and vegetables. These issues clearly blended together for her, heightening her sense of powerlessness. She described trying to just "cut out"—which I take to mean "stop participating in the system"—and concluded that she must try to "live as well as she can." But the process was tiring for her.

Phrases that Bygdabyingar used in connection to the topic of climate change—"we must take it as it comes," "we must try to live as well as we can," "it's just a joke to get involved," and "we can't do anything about it"—indicate a degree of profound powerlessness associated with this subject.

Feelings of Guilt

Thinking about climate change was also difficult in Bygdaby because it raised feelings of *guilt*. Members of the community told me they were aware of how their actions contributed to the problem, and they felt guilty about it. Recall the earlier interview passage where Eirik described the difficulty of both living by his conscience and living a normal social life in his community: "It is very . . . I think it is a bit problematic. I feel that I could do more, but it would be at the expense of, it would perhaps create a more difficult relationship between me and my children or my partner and in general. It really isn't easy."

Guilt was also connected to the sense of global warming as an issue of global inequity: Norwegians' wealth and high standard of living are intimately tied to the production of oil. Given their high newspaper readership and level of knowledge about the rest of the world, Bygdaby community members were well aware of these circumstances. This understanding contrasted sharply with the deeply ingrained Norwegian values of equality and egalitarianism (Jonassen 1983; Kiel 1993a), thus raising feelings of guilt.

It is also relevant that Eirik expressed his sense of guilt in the context of social relationships. He described how his connection to others in his community made taking action difficult. I explore this notion further when I examine the pressures of social norms in a homogenous society in chapter 4. Although privileged people around the world experience this contradiction between their wealth and the poverty of others, there is a particular force in the way in which these issues come together for people in Bygdaby and in Norway in general. High levels of access to information, high levels of acceptance of the information, a strong tradition of value for social welfare and the environment, and current wealth and economic interests come together with force. As the high school girls mentioned in chapter 1 stated, highlighting their discomfort with global inequality: "It's practically not right to sit here and just get, get, get; that's what we do, at least here in Norway; it's totally awful really."

Eirik similarly described the use of cars in his family and how the amount of driving they did negatively impacted his conscience:

Yes, if you take for example this with cars, we drive a lot of cars—in my family, that is. We go on vacation and we go shopping, and my partner drives to work every day. And I often drive up here [his office] myself. It gives us flexibility and so forth. And then we experience . . . we don't like it. We feel that we must do it to make things work in a good way, on a practical level, but we have a guilty conscience, a bit of a guilty conscience. We talk here about collective transport, but you know it really doesn't work. There are so few available that you must use a lot of time. Then you would have to give up a lot of things that you would like to do. So we are really hoping for some kind of other solution. For a car that is better, [that] doesn't create the same problems when we use it as what is happening now. Because I believe that we here, the way we live in Norway, we can't just stop using cars. So we have to hope that we will get a car that we can drive with a good conscience, (pause) a relatively good conscience.

Norwegian sociologist Ann Nilsen's interviews with Norwegian young people on climate change and their sense of the future contain similar expressions of both powerlessness and guilt. In an interview with a 23-year-old young woman in the study, the respondent offers her reflections in reaction to environmental problems and the third world (she had just mentioned climate change in the preceding passage):

It's terrible to think of, that we live so well while others live in such miserable circumstances. Of course it's very good to have a comfortable life . . . I enjoy it . . . but I feel so bad about the others, the rest. I have a guilty conscience, that's why I try not to think about it, keep it at a distance. . . . I still think these are important matters, but it's as if I can't make myself be concerned all the time, not any more. . . . Terribly important these matters, but I don't feel involved in a way, don't want to get involved. There are so many things to care about, so much information, we know so much about the connections between things in the world, in a way you are obliged to understand and to care. I suppose that's why my family has become more and more important to me, my everyday life, that which is near. (Nilsen 1999, 184)

This woman describes a fair amount of awareness of world problems and significant concern. She thinks "it's terrible to think that we live so well while others live in such miserable circumstances"; she knows "so much information . . . so much about the connections," yet she retreats from these thoughts, "keeps them at a distance," and "doesn't want to get involved." In fact, this woman says explicitly that it is because of her guilty conscience that *she tries not to think about these problems.* Note the connections between feeling and thinking. Fear of guilt and the attempt to maintain a "good conscience" ultimately cannot be controlled by not thinking about climate change. One remains aware beneath the surface that something is not quite right.

Fear of "Being a Bad Person": Individual and National Identity

A related concern with awareness of climate change is the threat it implies for individual and national identities. Although coming from a different tradition, social psychological work on identity complements work on emotion and cognition. The Norwegian public self-image has included a strong self-identification of being environmentally aware and humanitarian (Eriksen 1993, 1996). Norwegians have been proud of their international leadership on a number of environmental issues, including climate change. Stereotypical characterization of Norwegians describes a simple, nature-loving people who are concerned with equality and human rights (Eriksen 1993, 1996; see also chapter 5). Yet Norway increased production of oil and gas threefold in the ten years preceding my study of Bygdaby. Expansion of oil production in the 1990s contributed significantly to the already high standard of living, making Norway one of the countries in the world that has most benefitted from fossil fuels. In 2001, Norway was the world's sixth-largest oil producer and the world's second-largest oil exporter after Saudi Arabia (Norwegian Ministry of Petroleum and Energy 2002). Information about climate change—including Norway's inability to reach Kyoto reduction quotas, increased petroleum development, and participation in the Umbrella Group—makes for an acute contradiction between traditional Norwegian values and self-image, on the one hand, and the present-day economic situation in which high electricity use, increasing consumption, and wealth from North Sea oil make Norway one of larger per capita contributors to the problem of global warming, on the other. *Bygdabyingar* were widely aware of this issue. Øystein commented, "But what we have managed to do inside our own nation, it is clear that we have a ways to go still before we can keep the goal we have set, that which came from the international agreements. For example, with carbon dioxide we haven't managed very much; we have managed a little, but not so very much. It is, you know—we haven't managed, done very well, to solve that problem in Norway, I don't think."

Global warming was difficult to think about because it was threatening to individual and collective senses of identity and raised questions about whether people were "good," both individually and collectively. For Norwegians, information on global warming contradicts their sense of being environmentally responsible. And as a problem generated by wealthy nations for which people in poor nations disproportionately suffer, knowledge of global warming also challenges Norwegians' and Bygdabyingar's sense of themselves as egalitarian and socially just. In his work on denial, Cohen describes the scenario in which "you see what is happening, but you refuse to believe it, you can't 'take it in.' If these apparent facts and their manifest interpretations were true, this would seriously threaten your sense of personal and cultural identity" (2001, 24).

Information about Norwegian contributions to global warming disrupt the collectively created Norwegian sensibility and moral order described in chapter 1. During a conversation about global inequity, one of the many issues linked to climate change, Mona, a mother in her thirties, commented: "It is unpleasant to think that we in Norway hold others at a distance so that we can have things nice and good in Norway."

Torbjørn, a man in his early thirties raised in Bygdaby but who now lives in another town, commented: "We have a sense of ourselves as very good at giving money, but we are not as good as we think we are. People watch the TV program [TV Aksjon, a national fund-raising effort for a different global humanitarian cause each year] all day in the fall and see how much money people are giving, and they feel that we are generous. But when you compare it to how much is spent on a weekend on alcohol or for all these fireworks at New Year's. . . . When you are going to make a self-image, you choose positive aspects. Nobody chooses negative ones."

Svein, a friend of ours in his thirties who was a schoolteacher from a nearby community, told me about some of these contradictions as he sat at our kitchen table one evening:

Svein Norwegian schoolchildren have learned that they are not racist. That this happens elsewhere. But the Sami [an indigenous group in Norway] were treated terribly; up until the 1970s, they were taken from their homes, given schoolbooks in Norwegian. Of course, they couldn't read them. They were not helped. So there are many Sami people my age who are illiterate.
Kari It was the Norwegians who sailed the slave ships.
Svein Yes. We are proud of our sailing and shipping history, but we don't talk about that.

Emotion, Cognition, and Political Economy

Even in the face of such a highly emotionally charged problem as climate change, the emotion factor has long been missing from academic explanations for the public's nonresponse to the problem. Why and how might emotions matter? Emotions affect cognition in a variety of ways. As noted earlier, Lifton describes how fears of powerlessness or of being weak can prevent people from thinking about serious issues: "The degree of numbing of everyday life necessary for individual comfort is at odds with the degree of tension, or even anxiety that must accompany the . . . awareness necessary for collective survival" (1993, 108). Emotional needs and desires, Lifton tells us, influence what is and what is not acceptable to think.

Here we can also draw upon social psychology to expand our understanding of the possible significance of emotions for social inaction. Research from social cognition, sociology of emotions, and sociology of culture illustrate that people do not absorb in a direct, linear manner all information to which they are exposed. Rather, both individual and social processes operate in the organization of information that people hold in their minds. Paul Slovic (2000) describes how *affect*, or an association of "goodness" or "badness," is linked to judgments and decision making, including those involving environmental risks. Morris Rosenberg (1991) has examined emotional motivations for different interpretations of reality and the way moods and emotional states affect how people store information. Rosenberg describes how people self-regulate emotions because of a preference for emotions that enable them to get what they want out of life (instrumental) and because some emotional states are more enjoyable than others (hedonic), some are socially appropriate, but others are not (normative).

We heard from Bygdabyingar about their guilt and fears of being bad people. These emotions are closely linked with identity, an area that has been the focus of much research on cognition. Social psychologists Victor Gecas and Peter Burke describe the importance of our self-concept to the process of thinking: "the self is not simply a passive sponge that soaks up information from the environment; rather it is an active agent engaged in various self-serving processes," and "perception, cognition and retention of self-relevant information are highly selective depending on whether the information is favorable or unfavorable to one's self-conception" (1995, 50–52).

But through what process can social psychological needs actually influence perception? Feeling powerless, guilty, or concerned about the future might have been unpleasant experiences for Bygdaby community members in 2000–2001, but what exactly could they do about these feelings? Social psychological work on cognitive dissonance describes the needs people have to feel good about themselves and to feel that they can affect the world around them, both of which influence perception. But how exactly does this happen? As Rosenberg tells us, because emotions are difficult to control directly, "the main way of controlling one's emotions is to exert control over one's thoughts" (1991, 130). It would seem that people engage in a number of distortions and deceptions in order to maintain valued self-conceptions. Notice the links between emotion and cognition in Gecas and Burke's explanation that "people's self-conceptions are valued and protected and that a low self-evaluation (on criteria that matter) is an uncomfortable condition which people are motivated to avoid. This may occur through increased efforts and self-improvement or (more typically) through such self-serving activities as selective perception and cognition, various strategies of impression management, and restructuring the environment and/or redefining the situation to make it reflect a more favorable view of the self" (1995, 47). Of course, as Gecas and Burke note, conflicts between actual and desired views of self can also lead to action in the world. If Mona found it unpleasant to be aware of her role in global inequality, she could have "fixed" this problem by becoming socially active. Or she could have engaged in one of many options of how to restructure her perceptions. The fact that Mona found it unpleasant to think that she or other Norwegians might "distance themselves from others" might have made it less likely that she would want to pay attention to the situation that caused this unpleasant feeling.

Social psychological work applies to other emotions besides guilt. Recall that the topic of climate change raises feelings of helplessness. Yet Ellen Langer notes that "since feelings of inefficacy are undesirable and depressing, people may engage in distortions of reality and operate under the illusion of greater personal control and efficacy than they really have" (quoted in Gecas and Burke 1995, 48). Individuals may block out or distance themselves from certain information in order to maintain coherent meaning systems (Gecas and Burke 1995), desirable emotional states (Rosenberg 1991; Meijnders, Midden, and Wilke 2001a, 2001b), or a sense of self-efficacy (Gecas and Burke 1995; Bandura 1997).

Rosenberg's work on thinking as a technique of emotional self-control moves us toward a typology of options for controlling these unpleasant thoughts. The "denial" we have been looking at might seem so far to be merely in the domain of the individual. But at least some of the reasons for ignoring an issue are related to awareness of one's privileged position in the global economic order. Troubling emotions are troubling due to social context.

Emotion Management and Collective Strategies for Shifting Attention
Social structure (hence, sociology) is relevant for the study of denial because it draws our attention to the political dimension of emotions such as guilt. But it is not only the reasons for denial that are socially structured. Sociology also matters for the process of ignoring. Although the social psychological studies mentioned earlier provide part of the

picture of how emotion and cognition are linked through processes such as selective attention, we can also look to Arlie Hochschild's (1983) powerful theory of emotion management, which points to the importance of cultural norms and gives more details of the relationship between thinking and feeling. It is important to understand that sociologists think about emotions differently than either psychologists or the lay public. Whereas many nonsociologists regard emotion as part of what distinguishes an individual from society, sociologists of emotion describe emotion as deeply embedded in and reflective of both social structure and culture. Indeed, emotion can be understood as one of the main ways that social structures are reflected in our personal lives: "Many of the feelings people feel and the reasons they give for their feelings are social, structural, cultural, and relational in origin" (Denzin 1984, 53).

Sociologists of emotions describe *emotion norms* that prescribe the socially appropriate range, intensity, duration, and targets of feelings in different situations (Hochschild 1983). Emotion norms set the standard for what an individual "ought" to feel in a given context. Emotion norms apply to how people experience and respond to environmental problems. How scared should you feel when you see that the front page of the *Bergens Tidende* has a story on global warming? Also important are *display rules* that regulate the range and intensity of appropriate emotional expression in different contexts. How much of the fear one feels can be expressed? Emotion and display rules vary by contexts (such as public versus private spaces). And, of course, individual actors do not always feel or show the appropriate emotions. When what a person feels is different from what they are supposed to feel, they may engage in some level of *emotional management* (Hochschild 1983; Thoits 1996). Although the act of modifying, suppressing, or emphasizing an emotion is carried out by individuals, emotions are being managed to fit social expectations, which in turn often reproduce larger political and economic conditions. Arlie Hochschild's (1983) work on how flight attendants manage emotions in order to produce a pleasant experience for airline passengers (and thus increase income for the airline) is an excellent example of how the act of emotion management may link individuals, cultural norms, and political economy. In the case of global warming in Bygdaby, emotions that were uncomfortable to individuals were uncomfortable not just because they reflected a bad situation, but also because they violated norms of social interaction in the community. Thus, in addition to the social psychological explanations for nonresponse offered earlier, people also block out or distance themselves from certain information in order to follow norms of emotion. And in the case of the *Bygdabyingar* at least some of these emotion norms in turn normalized Norway's economic position as a significant producer of oil.

I place emotions and emotion management centrally in the process of denial. Emotions and emotion management matter for climate change because if the emotional states associated with thinking about a topic are uncomfortable or socially unacceptable, a person may not make the associated cognitive link. In Bygdaby, avoiding the unpleasant and socially unacceptable emotions associated with climate change was best achieved by simply not thinking about the topic. But even this statement is inadequate to explain what is going on. How does one not think about something that is important? It takes work to ignore the proverbial elephant in the room.

Because the primary way to control one's emotions is by controlling one's thoughts, the study of emotion management techniques draws on research at the intersection of theory on emotion and cognition (Hochschild 1983; Thoits 1996; Rosenberg 1991; Jasper 1997). Emotion management may take the form of either "surface" or "deep" acting (Hochschild 1983). When people pretend to feel an emotion they do not feel, they are doing what Hochschild calls "surface acting." With surface acting, the actor knows that she is trying to act according to cultural etiquette but that her own feelings are different. Emotion management becomes more complicated in the case of deep acting. In deep acting, one manipulates one's emotions to fit the social norms. Individuals talk themselves into the expected or "normal" response, usually by redefining a situation or shifting their focus. Thus, much of emotion management is cognitive. Furthermore, in the act of redefining a situation or shifting their focus, actors usually draw on existing cultural scripts. Through these existing scripts, we see the link to social structure.

In the case of deep acting, the boundary between individual and society becomes blurred, and as Hochschild describes, "the very act of managing an emotion is part of what the emotion becomes" (1983, 11). Through deep acting, individuals fit their emotions to the appropriate norms of their social position (i.e., gender, race, class, sexual orientation). Thinking can lead to different emotional states; thus, managing thinking is a central means to managing emotion (the emotion management techniques described by Hochschild in deep acting use thought control as a means of emotion control). As a result, if the emotional states associated with the information in question are not comfortable or considered appropriate, a person may not make the connected cognitive link.

But to say that there are unpleasant emotions associated with global warming is not enough to explain the lack of social movement activity in Bygdaby at the turn of the twenty-first century—especially considering that such emotions can also serve as the impetus for social action. I am interested in looking more closely at denial as a social process. How exactly might the presence of these troubling emotions put a damper on engagement? Despite the recent advance in the interest in emotions within social movement scholarship, little work has been done on the role of emotions in *non*mobilization (Goodwin, Jasper, and Polletta 2001). People in Bygdaby expressed emotions such as concern that would seem to motivate action. What, then, is the mechanism by which these same emotions might work to mute social action? Taking full advantage of this question requires not only that we keep in mind links between emotion and cognition, but that we employ an embedded sociological understanding of emotions. As we proceed to answer the questions of *why* and *how* people avoided emotions associated with climate change in Bygdaby, we move toward a picture of the social world that situates social action across the micro- and macrolevels, linking personal emotions and political economy.

On the one hand, the emotions people in Bygdaby felt provided much of the reason why they preferred not to think about climate change. But if we stop the story here, we would miss the bigger picture of how emotions are connected to social structure, of how denial is a socially organized process. In a sense, we would miss all the sociology. In the next chapter, I examine how emotion management is a central aspect of the process of denial, which in this community was carried out through the use of a cultural stock of social narratives to achieve "thought prevention," "perspectival selectivity," and "selective interpretation."

Throughout this chapter, it has been my aim to shift the discourse from one in which knowledge and caring about climate change are in short supply to a view whereby knowledge and caring are present but problematic and thus actively negotiated. Whereas this chapter has addressed *why* people failed to respond to climate change, chapter 4 examines *how* nonresponse is collectively produced through social interactions. We will see how the presence and management of unpleasant and troubling emotions associated with global warming worked to prevent social movement participation in this rural Norwegian community. In so doing, my work shifts from a focus on the information deficit model, in which information is the limiting factor in public response, to a focus on the importance of social context and political economy and the centrality of social interaction and emotion to both the cultural production of climate denial and the reproduction of power more generally.

4

The Cultural Tool Kit, Part One: Cultural Norms of Attention, Emotion, and Conversation

There were many things that we felt that we should have done, but we did like the others anyway, right. Because you have to fit into normal society. It is difficult. You can't just sign yourself out, right? If you did everything entirely ideal, you would be an outsider in the society, and then you wouldn't get anything done either.
—Eirik, Bygdaby community member

I have now described both the invisibility of climate change in Bygdaby as well as what people told me about why they did not want to think about it. But we must remember that the word *ignore* is a verb. Ignoring something—especially ignoring a problem that is both important and disturbing—can actually take quite a bit of work. So how did people manage to ignore this disturbing reality in their everyday lives? Questions about how people "create distance" from information on climate change and "hold information at arm's length" seem absurd if we take the everyday world at face value. But collectively constructing a sense of time and place, a sense of what is and is not appropriate to pay attention to or feel, is an important social and political process. In such constructions, we see the intersection of private emotions and the macrolevel reproduction of ideology and power.

Talking about Climate Change: A Bit off Topic

Local Conversation Norms
Political theorists such as Hannah Arendt (1970, 1972) and Jürgen Habermas (1984) tell us that when people get together and talk, a number of important things happen. Conversations are the site for exchange of information and ideas, for human contact, and for the building of community. They are also an important site for the creation of

collective meaning-making and reality (Giddens 1991; Gamson 1992; Eliasoph 1998). Conversations can help people understand their relationships to the larger world or can obscure those relationships. They can engage the sociological imagination. Conversation can also do the opposite. Yet conversations are also governed by rules called "conversation norms" that shape what is acceptable and not acceptable to say in different contexts (Eliasoph 1998). They are, so to speak, the rules of the game. By paying attention to conversation norms, we can begin to see the contours of social structure in private life and the links between political economy and interpersonal interaction. We can begin to see the contours of socially organized denial.

One factor that made it difficult for members of Bygdaby to think about global warming at the turn of the twenty-first century was that although they knew about climate change, there were few social or political spaces in which it was considered a relevant or appropriate topic for serious discussion. Here I describe conversation norms in five relevant settings: casual settings for small talk only, political settings, educational settings, social gatherings, and interviews. In chapter 2, I described the types of conversations that took place in these different social spaces. I refer to this material as I revisit each setting, here focusing on the norms that were operating in each space. In political meetings, for example, participants were to follow preset agendas. Topics that were not already identified were difficult to raise. Climate change was not considered a local issue and therefore did not appear on meeting agendas. In educational settings, however, instructors struggled to balance the task of teaching information against a fear of overwhelming their students. Within social and recreational settings, norms dictated engagement in "light" conversation only so that everyone could relax and have fun.

Small Talk

In Bygdaby, one of the few conversational spaces where climate change was raised relatively frequently was in casual conversation as "small talk." Here are a few examples taken from my research field notes:

An older woman got on the bus at a stop by a store at the crossroads. She approached the bus carefully, her arms filled with shopping bags. Despite the fact that it was mid-January there is no snow on the ground. As she paid her fare, she remarked to the driver, "This global warming is a good thing!"

A week later I sat at the bus stop with two women in their forties. We had a short wait and the two of them were chatting about the weather. One remarked to the other, "Things are so strange, you know, it makes me wonder about climate change."

A month later it was finally snowing in Bygdaby. I stood at the train station waiting for a taxi with skis in my hand. Outside it was snowing hard. I ran into Lene, the newspaper editor I had interviewed earlier that year. After we greeted one another, she noted with satisfaction, "Now you have yourself a real Norwegian winter [rather than the warm winters people have in most other nations]."

As in these examples, small talk occurred in spaces where people met in passing, while waiting for a bus or walking along the street. Not only was the small talk meant to fill a short time space, but on such occasions people were often seeing someone they hadn't seen for a while. Launching into political discussion would seem inappropriate. Thus, although the presence of climate change as a topic of small talk worked to keep weather and climate change in people's awareness, it did not allow for displays of deeper feeling or generate the kind of more serious analysis that could lead to a sense of what might happen in the future or what could or should be done now.

Conversations on the Political Scene

A second conversational setting, one that is especially relevant for the discussion of an environmental problem with major social and economic ramifications, is talk that is explicitly political. In chapter 1, I described the relatively high level of political activity in Bygdaby. At least eight political parties were particularly active, holding strategy meetings and putting forward candidates for local election. I met with political leaders from the Center, Socialist Left, Labor, Communist, and Christian People's parties and attended half-a-dozen meetings of the largest and most active party, the Labor Party, and a number of meetings of the city council subgroup on culture and environment. Yet, as described in chapter 2, climate change was never discussed in any of the specific Labor Party strategy meetings I attended or in the meetings of the city council's culture and environment subgroup. Climate change was, however, a topic of inspirational speeches given at the annual meeting of the Labor Party and by the prime minister and the king on New Year's Day—events that were highly visible and inspirational but not part of day-to-day political practice. In chapter 2, I described what did get discussed in these meetings, such as downtown zoning laws, new areas for huts, and budget decisions. Why were these topics considered legitimate, whereas the local economic and cultural ramifications of climate change were not? What were the conversation norms in these meetings? Did these norms connect in any way to the lack of talk about climate change?

The conversation norms in these political meetings followed a norm of focusing on local events in isolation. This framing made the unusual weather of the season appear more anecdotal rather than connecting it with larger global patterns, and it thus created a sense of climate change as distant, happening somewhere else, and less relevant. In the absence of a well-developed sociological imagination that connects individuals to society and the local to the global, private or local troubles look merely personal rather than political, and their consequences seem less significant. Although such constructions of the local feel natural, they are a way of "not seeing" that has powerful implications for social action. Susan Opotow and Leah Weiss describe in their work on denial and moral exclusion of environmental conflict the phenomenon of "temporal containment of harm" in which we may assert that exposure to or injury from harms is an isolated, unlikely event rather than routine or chronic or both (2000, 481). They categorize temporal containment of harm as a form of "denial of outcome severity" or the concealment of environmental harm based on selective distortions.

Opportunities for the topic of climate change to emerge in political meetings were also constrained by the expectation that participants were to follow preset agendas. For example, at one point in a meeting of the Labor Party about how to improve party membership, members began a more visionary discussion of what they felt community members cared about and how to create dialog with the community. It was one of the more lively and interesting meetings of this group that I attended and one of the few conversations I observed in which issues beyond the immediate valley were discussed. In this conversation about how to recruit more party members, one man, Svein, felt that people were more concerned with local things ("opptatt av nær ting"), but Andre and Gurid raised larger global issues. Andre began the debate when he spoke passionately about the significance of several global issues, especially AIDS. However, Jon, the chair of the committee, said, "Bring it closer to home, to the situation here in Bygdaby," and Svein reiterated, "Local things, people care about local things" ("Nær ting, folk er opptat av nær ting"). Gurid interjected, "But people are concerned about national things and international things, too." To which Jon replied, "People are concerned about the schools and whether older people have the care that they need." In this conversation, party members reconstructed a sense of emphasis on the local. The chair of the meeting repeatedly enforced this norm explicitly by reminding the speaker to "keep to the topic of membership." Somehow the larger issues and discussion of what people cared about and how to create more dialog in the community were not perceived as connected to the defined goal of increasing membership.

This use of what is "local" to patrol the boundaries of appropriate action has also been found in the United States. In her study of "the production of apathy," Nina Eliasoph noticed that the people she spent time with focused on issues that seemed "doable" and "close to home." She describes the phrase "close to home" as indicating not those things that are necessarily closer in physical space, but those issues that people feel a degree of control over. Eliasoph notes that some respondents defined very physically distant issues as "close to home" because they were "doable" and other issues, such as the nuclear battleships that were in their front yards, as distant because they felt powerless to do anything about them. It appears that constructing climate change as a distant issue, in either time or space, may have served this purpose for *Bygdabyingar* as well.

Educational Settings: Be Optimistic

In educational settings, a site in which one might expect information about climate change to be discussed most, pressures to be optimistic limited open discussion of the seriousness of the problem. When talking about difficult topics, educators described balancing personal doubts and deep feelings of powerlessness with the task of sending a hopeful message to their students. All three of the educators with whom I discussed climate change raised this point. Arne, the teacher at the local agricultural school, told me, "I am unfortunately pessimistic. I just have to say it. But I'm not like that toward the students. You know, I must be optimistic when I speak with the students."

Arne's use of the phrase "you know" highlights the sense that this reality, this need to be optimistic with students, is taken for granted, incontestable. This "need to be optimistic" reflects the unspoken fact that the material about climate change that these teachers were trying to communicate brought up the disturbing emotions I described in the previous chapter. In order to communicate this information, the educators had to fit it within this social norm. Geir, the environmental activist I spoke with in Oslo, described using dramatic information to convey the seriousness of climate change but emphasizing solutions at the same time:

In the first place we have a media society, so if you don't have something that is fairly strong then you can't get your point made. And often then to try to scare people becomes an essential way to get your point made. But I also believe that it is important to speak the truth. And there are things that are quite frightening. But we try, always in the same sentence as we talk about serious problems, to say that there are solutions and that these solutions are not particularly difficult.

Another teacher, Kristine, designed environmental education programs for elementary-age students in the Bygdaby community. During the interview, she described how she was actually cutting back on an emphasis on environmental degradation at the local level because in telling young people about environmental problems, she felt that she was "taking nature away from the children." Her use of the phrase is based in the very common Norwegian sense of nature as a refuge and as children's birthright.

Intimate Spaces and Social Gatherings: Shelter and Recuperation
Another possible space for conversations about climate change is the informal social gathering. The study of conversation analysis, including work on conversation norms, grows out of Erving Goffman's (1959, 1974) important theory of ethnomethodology. Goffman tells us that human social behavior is highly performative. Perhaps here, in what Goffman would call a more "backstage" setting where people are more relaxed and pressures to perform are eased, the troubling topic of climate change can at last be raised. What were the conversation norms in such spaces in Bygdaby, though? In my field notes, I wrote:

It was four in the morning at the staff party of the bar where I worked. Several dozen young people were spread out on couches and chairs in the low light of the club. Pop music came from a CD player, candles on the tables were lit. Most of those present had drunk three or more drinks. All were visibly relaxed, and a few quite drunk. Gunvar and I were talking about the ways that people act differently in different places regionally in Norway, including drinking behavior. Gesturing at the group of drunk people spread on couches around us, he said, "Det er så mye vondt i verden . . . det er så mye vondt i verden at det er viktig å være her i lag" (There is so much pain in the world . . . there is so much pain in the world that it is important to be here together).

Gunvar's comment emphasized the importance of friends "being together," sticking together in tough times, and the important purpose of such settings in parties to provide a space of shelter from the troubles of the outside world. He implied that alcohol use is understood as a means of coping with these difficulties. This was a place where much of the isolation and distance that resulted from everyone's "holding it all in" was broken down. Might this space then be one where private fears regarding future climate scenarios could be voiced? But here as well there were pressures not to be too negative. Friends listened to stories from each other's lives and gossiped about local happenings. Part of the expectation of these social spaces was that people needed a place to come together and relax, to recover from the challenges of the wider world.

Because these spaces were constructed as shelter or refuge from the challenges of the wider world, serious topics were generally avoided. But they occasionally did come up. With the aid of alcohol and the permissive sense that "anything goes" in spaces where alcohol was present, people did sometimes say more serious things that were on their minds. It was also easier to meet people in spaces where alcohol was present. I certainly had a number of open and candid conversations with community members about life and the state of the world in such settings. I described earlier how one evening Finn shared his concerns about having children because he did not know what kind of world his grandchildren would have. These confessions of normally private fears were important. Yet fears that were shared with friends late at night in half-drunken states (or thoroughly drunken states, as was not uncommon) were not really fears that got publicly expressed. They remained backstage and private. Although perhaps important to kindle ideas, these feelings were not yet being translated into political action.

Like the bar, the home in Norway is also viewed as a sheltered space for recuperation (Gullestad 1997). Yet in intimate spaces in the home, among families, people told me they sometimes held back from speaking their minds because they wanted to protect their loved ones from troubling feelings. As described in this conversation with Øystein from the last chapter, the local politician who was active on environmental fronts, parents in particular expressed reservation about talking about climate change with their children. When I asked him whether he spoke about climate change with his preteen children, he replied: "Not so much, more sporadically. It isn't often, but of course they have this theme at school, the kids. They have it at school of course, so they hear it of course, they are aware of it. They have it at school so that the kids will have a good relationship with the environment and so forth, but I don't talk so very much with my kids about it. Not so very much. . . . We don't talk about it much."

Talking about Climate Change in Interviews: Deeper Feelings but a Bit "Off Topic"

Finally, as I described in chapter 2, I noticed that when I raised the issue of climate change in interviews, it was difficult to keep to the topic. At least in Norway, the sense that climate change was important was almost universally expressed. Of the 46 people I interviewed, only one man said that he didn't think that global warming was happening, and one other said that he wasn't that concerned. Yet once an interviewee and I finished discussing the local weather itself, we very quickly hit a zone where there was suddenly "nothing to talk about"—that is, people had general information but were not comfortable thinking or speaking at length about potential consequences of these changes, future weather scenarios, or what, if anything, might be done to reduce carbon emissions. Community members were able to provide a great deal of information about weather patterns. And they realized what was at stake in the situation—they told me so specifically—yet these acknowledgments were smoothed over by phrases that made light of the situation, a shrug of the shoulders, a wave of the hand, which seemed to indicate "What is there to do?"

Was the common sense understanding of the issue, as Åse put it, that "it's just a joke to get involved"? I found this pattern of knowing and not knowing, seeing but not seeing, similar to Nina Eliasoph's description of the way in which "in interviews volunteers managed both to see the problems and to turn away from them, describing them in lavish detail while simultaneously avoiding them" (1998, 70). Why did Øystein, a man deeply involved in local politics, tell me on the one hand that climate change was the single-largest environmental problem facing Norway and the world, that the problem was so significant that "it is possible that in one hundred years the environment will be damaged to the point that it isn't possible to live on earth anymore," yet on the other, did not discuss this issue with his preteen children? How could Hilde tell me one minute that it was possible that climate change had gone so far that it could not be turned back, yet in the next sentence optimistically add that it would be "exciting" to see what would happen, as though we were discussing the outcome of a local soccer game?

Why were people able to make small talk about the weather but unwilling or unable to talk about climate change in a more serious way? These gaps and contradictions were evidence that notions of "knowledge" and "caring" regarding climate change cannot be simply taken at face value, as occurs with a decontextualized survey, but instead need to be problematized through a more nuanced ethnographic interpretation. In this context, Hilde's use of the term *exciting* worked simultaneously to put a positive spin on the situation and to normalize it. She was also simultaneously constructing her status as a bystander, someone on the sidelines who will be entertained rather than someone who will be impacted. These forms of spin are part of Cohen's notion of interpretive denial, of what Opotow and Weiss call "denial of outcome severity," and of what Eliasoph considers a strategy of maintaining conversational control.

Taking these various conversation norms together, we can see that in Bygdaby global warming was most often mentioned in a superficial or general way, as a topic of small talk. Though it was raised in more public, inspirational political speeches, I observed no discussion of it in daily political work. These contrasts between what gets expressed as private daily concerns and what counts as public political talk are again similar to Eliasoph's observations on political apathy in the United States. Eliasoph notes, "The people I met did sound as if they cared about politics, but only in some contexts and not others. They did not just think everything was fine as it was, but there were too few contexts in which they could openly discuss their discontent. Most of the time intimate, late night, moonlit conversations were the only places other than interviews where that kind of discussion could happen. In group contexts, such discussion was almost always considered inappropriate and out of place" (1998, 7). This is not to say that there were no in-depth political conversations about global warming happening in Bygdaby, but rather that people were having such conversations in spite of cultural norms, not because of them, and those norms still limited the extent of such conversations, even in spaces where such norms were eased (see table 4.1). We move next to a discussion of the emotion norms that intersect with and underpin many of these conversation rules.

Table 4.1
Conversation Norms and Climate Change in a Variety of Settings in Bygdaby

Small talk	Weather, casual comments
Educational settings	Be optimistic
Social contexts	Time to relax, take shelter from the wider world
Political settings	Local focus, preset agendas
Interviews	Deeper emotions expressed, yet with discomfort—still a bit "off topic"

"You've Got to Be Strong": Emotion Norms

Feelings such as guilt and helplessness are not only unpleasant to experience, but also inappropriate to reveal publicly; they are emotions that cultural norms bar from public expression. Emotion norms in Bygdaby (and in Norway in general) emphasize the importance of maintaining control (*beholde kontroll*) and toughness (*å være tøff*) and, for young people, of being cool (*kult*)—especially in public spaces. Adults, especially men and public figures, experienced pressure to be knowledgeable and intelligent. And just as there are political implications to conversation in terms of the development of awareness and construction of reality, control of emotion has deeply political implications. In this section, I describe three emotion norms that shape acceptable cognitions in Norway: being tough, being cool, and being smart. Table 4.2 gives an overview of the range of avoidance or normalizing strategies I found to be in use in Bygdaby with respect to global warming.

Maintaining Control and Being Tough

During my stay in Norway, I took a trip by train every month or so. One afternoon as I settled into my seat and opened the train's free tour magazine, I found a humorous advertisement that played on a prominent emotion norm, "being in control." On the inside cover was an ad for a cell phone. In the picture, a couple was sitting at a café. The image showed a man holding the cell phone out of sight from his date and sending a text message ordering flowers to be delivered to their table. The text of the ad said, "Helping you maintain control."

"Maintaining control" was extremely important in public spaces for both sexes. Because the concept of maintaining control was linked to control over emotions, it was associated with another key value of "being tough," which was in turn connected to the notion of independence. Knut, a community member in his fifties, explained to me how Norwegian men are expected to be tough, at least on the outside:

Table 4.2
Climate Change and Emotion Norms in a Variety of Settings in Bygdaby

Being cool	Care about boyfriends, not world problems
Being tough, maintaining control	Don't show feelings of powerlessness, uncertainty
Being smart	Have a good answer

Both Norwegians and Finns can be cold and hard. Hard guys. But inside themselves it's pretty soft, and that's why you get these ways of releasing it, because you get drunk. And then it isn't long before the tears come. . . . When you look at this, we Norwegians are maybe a bit introverted, and here in Bygdaby we have an expression, "to be a man." It means that you behave like a man; as an ideal man, you can say . . . it's things like, you can't stand to be different, you fit in, and you fit in, and you are tough, right. So you drank, really, before you were able to say what you really meant. And it is an attitude that, I think, maybe we can go back to history [for], to the Viking ideal. It was like, the way you should be. Strong, don't say anything, you should be able to take it all, take anything, you shouldn't show your feelings—just think of the Norwegian sagas. There is really nothing [in them] about love.

Here Knut explained the links between lack of emotional expression ("you shouldn't show your feelings"), saying what you really mean, and in fact saying nothing at all. Knut also referred to a frequent characterization of Norwegian men as introverted, shy, and socially a bit backward (see chapter 5 as well). Being in control is about "fitting oneself in" and about not drawing attention to oneself. In contrast to cultures where masculinity is linked, for example, to sexual prowess, here, as Knut described to me, the emphasis is on strength as emotional control (athletic accomplishments are also very important).

Knut then went on to tell me an old story about a man from Iceland, which shares a common history and culture with Norway. This man went to see the king, but he had a swollen foot. When he arrived, the king saw that his foot was very swollen and asked him why he wasn't limping. "Why should I limp," the man replied, "when both my legs are the same length?" The subtext here is the control of physical pain. As a man in this community, one should be able to endure pain. Readers will note the gendered nature of the story, and, indeed, these norms are especially strong for men. But women and men alike are expected to be tough and take charge.

Love in Norway is considered difficult to show, in part because it conflicts with the norm of independence. Norwegian sociologist Marianne Gullestad's ethnography of young women describes that as close as these friends were, "at the same time, they seldom talk about personal feelings of sexual love" (1992, 301). A female friend of mine from another community explained to me explicitly the importance of the norm around love for Norwegians. My friend went to great lengths to clarify that "if you are Norwegian, you don't just say 'I love you' to someone, even if you do love them. When you say [it], it will be very hard to say and therefore very special. It isn't like in the United States."

Although people in many cultures might describe it as "hard" to say "I love you," we can imagine here the difference in expression norms in Latin cultures, where vivid romantic pronouncements are regularly glamorized through music and poetry.

My friend's explanation and reference to the United States relates to another dimension of the local emotional culture regarding minimal public emotional displays. Here, too, Americans are held up as a negative example in explanations of this norm. Americans are described as superficial because, according to my friend, "they ask one another 'how are you' without really meaning it." Of course, like most Norwegians' view, my friend's understanding of the United States was heavily influenced by Hollywood movies. However, I heard similar comparisons to and statements about the United States repeated many times from *Bygdabyingar*, especially when I asked about cultural norms that were different to mine, such as why people did not greet one another on the street. It is common for people who know each other by sight but have never been formally introduced to avoid eye contact and pass each other on the street without greeting. Unlike in my own country, where I am likely to say hello to whomever I walk past, in Bygdaby even people who have known each other for years may not greet as they pass one another on the street.

My conversation with this young woman about expressions of love and constructions of superficiality offers insight into emotion norms and the sense of "keeping your feelings to yourself, don't be vulnerable." Vulnerability in Norwegian society is much more costly than in the United States because it involves breaking taboos. This pressure to be tough, in control, and, as a popular Norwegian folk song goes, "alltid ser lys på livet," always looking on the bright side, is also connected to the pressure to be optimistic or at least not to "break down" in public.

When two American tourists got onto a bus in the spring and announced with pleasure, "Oh, it's so nice and warm in here," I realized just how adjusted I had actually become to the norms of Bygdaby. These women were not the typical "loud, obnoxious American tourists." Their "loud" voice was well within normal volume for an American. Yet compared to the Norwegians on the bus, who were sitting still and upright in their seats, at most whispering, but in general otherwise not making a sound, the tourist's audible comment broke a social taboo. These women drew attention to themselves. Compared to *Bygdabyingar*, most Americans make eye contact with and smile more often at strangers, walk in a more relaxed manner, admit when they don't know things, ask stupid questions, and show a much wider range of feelings in public, be they pleasure, irritation, or sadness.

For example, the young women who did the school project on AIDS used the emotion norm of toughness as a standard to criticize other classmates who were unable to deal with disturbing information. In this passage from the school focus group, Trudi reframed Ingrid's assertion that schoolmates didn't care about world problems to mean that they weren't tough enough to deal with them:

Kari In any case, I think that it can be difficult to begin to learn all this information, that not everyone in the world has things as good as I do. Can you talk a bit about it, about how it went for all of you? As you began to learn about the situation of others?
Ingrid I think there are a lot of people who don't really care . . .
Kari That they don't think it is important?
Ingrid I think they have to learn a bit more, get a little closer in to the material before . . .
Trudi (Interrupting, with force.) They don't dare to get any closer to the material.
Siri It has something to do with that, when you have it closer to you. . . . So you feel so distanced that you don't experience it. But I have noticed that I have reacted a lot more when it comes to things like AIDS and contagious things.

Here, "not daring to look at the information" is a sign of weakness, a bad thing. From other statements by Trudi and the others about it being considered "uncool" to think about world problems, it seems that Trudi was framing their group's activities as better than the "not caring" of those who "don't dare" to look at information. She did so using the available normative pressure of "being tough."

Being Cool

Another emotion norm that shaped responses to climate change in Bygdaby in 2000–2001 was that of "being cool." As is true around the world, the importance of being cool applied particularly to teenagers. It was not cool to be too serious about the world. Teenagers who took on serious problems, such as the young women who worked on the AIDS project, had to face this pressure head on.

Trudi There is a bit of peer pressure, you know, like it isn't cool.
Mette It isn't cool to be concerned with . . . the world's problems.

Trudi That isn't how it is for us, but I think that it's like that for many. We who are already more active, we don't worry about whether it is cool or not. But there are a lot of others who are probably concerned about it.

As this passage indicates, being "cool" was important for young people and not always equated with concern for the wider world. Being cool was about boy/girl friends, parties, and having a good time.

In Bygdaby, the norm that one should be in control and decisive yet relaxed is enforced sometimes through the phrase *slapp av*, or "relax." People who appeared too serious might, for example, receive such instruction.

Being "Clever" or Smart

Especially for men and for adults who were in professional or leadership situations, there was a palpable pressure to appear smart or "clever," as *Bygdabyingar* would say. One man who had been politically active for many years responded to my interview questions as though he was taking a true/false quiz on stage, answering each question with a cheerful, brisk, and self-satisfied confidence. At the end of the interview, I had the sense that he expected to be congratulated on his performance. Another man, Geir, who was from Oslo, typified this characteristic of emotionless control in which the problems of the wider world were on the one hand disastrous and on the other hand did not personally affect him. He himself was not shocked by the information he gave the public on climate change; rather, he was the one who dispensed shocking information. In our interview, I attempted to understand these norms a little better by asking Geir questions about his own personal trajectory regarding his work. I asked him how he first came to be involved and when he had first heard about climate change. When he responded, his tone was impatient, as though it was a silly question:

Geir As long as I can remember, I have loved nature and animals; from quite far back I also have the, I think, quite natural feeling that I didn't want others to suffer unnecessarily. And when I then began to hear that the world was so different in many lands other than Norway. And then as a 14- or 15-year-old I began to sense that there was somehow a connection between our wealth here in Norway and the poverty in other parts of the world; it made a big impression on me. And then later there came new information about environmental poisons and still later about climate change. So it [my awareness] has been there the entire time, but gradually increased.

Kari So there isn't a particular point that you would trace it from?

Geir (Impatient again. Begins answering before I have finished asking the question.) No, no, it isn't that I found God or that something like that has happened; it was just gradual. (Here I suspect his impatience is due to my breaking a social norm in pressing him for personal, potentially emotional information.)

Kari Is there an environmental problem that hits you in a particularly personal way?

Geir (Again slower, more thoughtful.) All these problems are so connected. The greenhouse effect is an enormous threat for both poor people and biological diversity. But if I am going to pick out one point, it would have to be species extinction. On the one hand, there is suffering for people; on the other hand, there is suffering for animals, but worse than the suffering of individuals is the extinction of a species. And today it is known that 25 percent of mammals are threatened by extinction. . . . That is the most dramatic. But, of course, there could be an atomic accident that suddenly means that it is the most important problem.

Throughout the interview, Geir presented himself as fully in control. Despite my repeated direct questions about why and how he became involved and what issues mattered most to him personally, he did not talk about his personal feelings. Instead, he framed his work as a matter of acting in the rational way to bring around ecological sustainability, but not as something that came from any personal connection. He did not really answer my last question about whether there was an issue that he personally was more concerned about, but instead gave his evaluation of which problems were most important. His impatience also could have been signaling his own superior knowledge on the subject—that he, having figured it all out, was impatient with my silly questions. This particular conversation was one of the most extreme cases of the emotion norm of "maintaining control" that I experienced. And although both men and women expressed feelings of powerlessness and loss of control when thinking about climate change, the cases in which the self-presentation of maintaining control were most extreme involved men. Yet climate change is a problem for which nobody can have good answers to everything. Even if we know the latest figures on necessary carbon dioxide reductions or expected sea-level rise, there is inevitably more we can know about specific technologies, political policies, and the latest scientific research. Thus, in the face of this norm to be clever and in control, and especially given the emotions that climate change potentially raises, thinking seriously about it was constrained or even censored in Bygdaby.

Cultural Norms of Attention: The Construction of Time and Space

The normative delineation of our attention and concern is one of the most insidious forms of social control. Through a variety of norms of focusing we internalize as part of our "optical socialization," society essentially controls what thoughts even cross our minds.
—Eviatar Zerubavel, *Social Mindscapes*

From the perspective of sociology of cognition, people learn to think through socialization into different "thought communities" (Zerubavel 1997, 9). Events occupy our imagination, our conversations, and our hearts, producing the sense of what is near and far, significant and insignificant, personally relevant or irrelevant. Social norms of attention—that is, the social standard of "normal" things to think about—are powerful, albeit largely invisible social forces shaping what we actually *do* think about. Just as social norms of attention create the sense of what *is real*, they also work to produce the sense of what is *not* real, what is excluded from the immediate experience of normal reality. In Bygdaby, disturbing information was also held at a distance by following established cultural practices of what to *pay attention to*.

The residents of Bygdaby created both a sense of community and a sense of reality through what they collectively paid attention to and what they collectively ignored. Norms of attention set boundaries for awareness in terms of time and space. How far into the future or past was it normal to think? How close or how far should one focus one's attention? It didn't take long to notice that norms of attention in Bygdaby that directed attention to local happenings and to the past were at odds with the scales of time and space needed to conceptualize climate change, to make it seem real.

Scales of Time: The Presence of the Past
Daily life in Bygdaby, especially for the longtime *Bygdabyingar*, is marked by a pronounced sense of the past. Within Bygdaby, it is much more normal to think about what is happening in the community 200 years ago than to ponder what might happen 20 years in the future. In the

words of Kjell Magne, community members are "preoccupied with culture, it is completely true. . . . Here in Bygdaby we are really good at taking care of our culture and building on old traditions. We really want for things here in Bygdaby to remain as they are."

Work activities, in particular those associated with farm labor, are in many ways done as they have been for generations. There is a rhythm to life. Lambs are born in the spring, cows and sheep go to the mountains to graze in the summer, and sausages are made in the fall and winter. This is not to say that things have never changed. Cows are now milked with milking machines, and fewer and fewer people work the farms; once the activity of an entire family with servants, farming is now carried out by solitary farmers working in relative isolation. Changes have happened, yet seemingly everywhere throughout daily life connections to the past are interwoven—socializing around beer production in late fall, drinking from a wooden cup that has been part of a farm and its history for more than 200 years. Both traditional activities and physical artifacts make the community's past history "real," "close," and visibly felt in present time.

During winter, the recreational activity of skiing connects people to the past. Skiing is not only a practice with a very long local history, but also an activity that takes place in the *utmark*, uncultivated land (usually in the mountains where sheep and cows grazed in the summers). A trip into the *utmark* is thus an opportunity to review the history of the landscape. On one trip, I skied past a series of small markers, each with a name: Kvilesteinen (the Resting Stone), Raudalen (Raul Valley), and so forth. They are the historic place-names. When one skies in the high country, one often comes to an old *støl* (or *setter* in standard Norwegian), a mountain farm where in the past women and children lived all summer tending herds, milking cows, and processing milk into cheese and butter. Although old, these buildings have almost always been carefully maintained. On the group ski trips I attended, people talked about these places. We might stop at a *støl* for lunch or a short rest. Trip leaders shared information about who owned the property and maybe told a story or two about that person. When one skis into one of the many mountain huts operated by the volunteer Bygdaby Utferdslag, such commemorative culture is more formalized. One finds older photos of *Bygdabyingar* on skis and history books filled with such images and stories from the past.

The sense that the past is more real than the future is also created through the arrangement of physical space in the town, which I described in chapter 1. Physical monuments, traditional practices, and local institutions serve to orient the collective focus of the community backward in time. Throughout downtown Bygdaby, one finds dozens of *minnesmerker*, memory markers, that are monuments to past events—a stone cross marks when the village became Christian, another stone marker indicates the place where Russian soldiers camped during World War II, a sculpture downtown commemorates a local man who became a famous artist during the twentieth century. The fact that the past is important to the people is highlighted by the emphasis on preserving old buildings and learning about their histories. Not only is there an official public body dedicated to this task, but there are numerous volunteer efforts. One of the community groups that I spent time with during my stay in Bygdaby, the local Rotary Club, had recently produced a beautiful full-color guide to the history of the various monuments and *minnesmerker* of Bygdaby. Club members had gone to great lengths to take professional photographs and collect detailed historical information on the dozens of monuments and markers and historic grave sites found within this small community.

Although the presence of the past generations feels "natural" to community members, it is nonetheless being actively produced in both conscious and unconscious ways. Significant economic resources go into maintaining historic buildings and sites, thanks to the efforts of the political body dedicated to environment and culture. And community members produce the sense of the past on a more unconscious level by following unwritten rules about maintaining social relations—for example, maintaining the boundary between insiders and outsiders mentioned in chapter 1.

Environmentalists have described how Western societies' failure to think on a longer time scale is part of why we have created long-term environmental degradation such as nuclear waste. In contrast, the Iroquois Nation is reputed to make decisions from the perspective of how they will impact people living seven generations in the future. From a sociological standpoint, these issues concern the social organization of time. Although there are many reasons why thinking about the consequences of climate change is not part of daily life in Bygdaby, one of them is clearly the disjuncture between the sense of time as it is normally experienced and the sense of time necessary to observe climate change or make its consequences seem "real." Although this experience of the world appears and feels "natural" or "inevitable," perceptions of time are in fact socially produced.

Tradition is very important to the community. People prepare and eat traditional foods, wear traditional clothing on significant days, and keep to their traditional dialect. All this serves to orient the community's collective focus backward in time. It is not an uncommon remark that people from Bygdaby "har øyne i nakken" (literally "have eyes on the back of their heads"), meaning that they look more to the past than to the future. In 2000–2001, Jorn was an administrator with the city, and for him this emphasis on tradition and the past made it harder to do things in new ways or to do them differently: "People are so focused on things that happened twenty years ago. They tell you that you can't do this or that because twenty years ago that person did x or y." These observations about time, space, reality, and daily life are parallel to those made by Norwegian social anthropologist Ann Nilsen. Nilsen interviewed young people in Bergen about environmental problems and their sense of the future. She found that environmental problems based in the future were difficult for these young people to think about or imagine:

Incorporating the gloomier sides of existence, such as environmental problems and their effects into the scenarios made for their own lives ten years on, is near impossible for everyone. In a sense the problems are thought of, and talked about, with the same sense of distance as are dreams—they are related to the "unrealistic" aspects of thinking about the future. Even if one is fully aware that environmental problems in threat terms exist, somewhere, everywhere in fact, and even if one knows the possible consequences of the most threatening of these, they are very difficult to relate to as personal concerns, also in the future. (1999, 190)

Nilsen found it useful to divide her analysis of young people's conception of time into the categories of dreams, hopes and plans, and nightmares. Dreams and nightmares are characterized by a kind of unreality; plans are the most tangible; and hopes belong to a world in between. Future environmental problems belong in the category of nightmares: "Like nightmares disappear and seem distant in the light of day, the worst-scenario environmental disasters seem so abstract and distant that they cannot be thought of as 'realistic'" (1999, 190). Nilsen concluded that according to this perception of time, "in societies such as the contemporary Western world where thinking and attention span are aimed at the extended present, or the immediate future, environmental problems of the magnitude that climate change represents, for instance, will be difficult to find solutions to, also because of a general time horizon involving less attention to the long-term future" (1999, 176).

Memory is a function of socially organized salience. What people choose to remember and forget is a reflection of the needs of the present. Bygdaby and Norway as a whole are experiencing rapid social, political, and economic change. Whereas once *Bygdabyingar* worried about basic survival, now they face a new set of threats that range from the decline of the welfare state to increased radiation levels caused by deteriorating Russian nuclear facilities and the ecological chaos predicted by climate change. As the future begins to look problematic, some community members, especially the older generation, have changed their focus of attention and begun to look backward even more. Here it is appropriate to mention Eric Hobsbawm and Terence Ranger's ([1983] 1992) concept of *invented traditions*. Their work describes how what we think of as "traditions" may be quite recently constructed and furthermore are practiced today because they serve particular needs of present generations. In the face of uncertain futures and a confusing wider world, ideas of tradition and links to the past serve as an anchoring point in changing times, providing a sense of security that at least some people believe helps them to deal with the larger world.

Hiding under the *Nisselue*: The Social Organization of Space

Norms of attention mark not only the social experience of time, but also the proper spatial range of attention. Events that are "close" are considered more real and worthy of attention than those that are "far." Yet how close is close and how far is far? How is the social experience of place, this sense of the local, constructed?

It is through managing boundaries of spatial attention that people create the sense of "imagined community" that Benedict Andersen (1991) describes as forming a nation. Events or information relating to places or times outside the cultural sphere of attention can be classified as "unimportant" and tend to feel "unreal," "inaccessible" not "close to home." In Bygdaby, it is more normal to pay attention to events that happen in northern Norway than to events happening in Rome, which is roughly the same distance from southern Norway as the North is, because the latter occur outside national boundaries. Despite the physical distance, events in northern Norway *feel closer* than those in Rome or even than those in Denmark, which is actually much closer to southern Norway than northern Norway is.

Thinking about global warming obviously requires not only an ability to visualize the future, but also the ability to imagine events that are taking place elsewhere, such as melting polar ice caps or flooding islands in the South Pacific. In Bygdaby, an emphasis on the local is created through focusing on local events and history, maintaining a strong boundary between insiders and outsiders, reading the local newspaper, and emphasizing *lokalepatriotisme*, local patriotism. Soren, one the most active environmentalists in the community, described this sense of parochialism, which I jotted down in my field notes:

> As Soren put on his hat and gloves and prepared to step out into the cold May evening, he noticed the small red button that was pinned to the curtain of the house, a relic of the house's former occupants: "Stem Nei," it declared, "Vote NO!" [referring to European Union membership]. "That's just typical Norwegian," he commented. "Oh," I replied, ever eager for an opportunity to hear a local person's perspective on such matters, "how so?" "Well, the sense of just being focused on ourselves, kind of like hiding our heads under the sand, only we have a saying "to pull the *nisselue* [a typical Norwegian hat] down over your shoulders. It's like we don't want to know; we don't need to worry about what happens outside our borders." (Field notes, May 2001)

The *nisselue*, a type of hat worn when skiing, is an image that is highly symbolic for Norway and equated with earthiness, simplicity, backwardness, independence, and rural life (Eriksen 1993, 22–23). The phrase "pulling the *nisselue* down over one's eyes," similar to the expression "hiding one's head in the sand," is commonly used to refer to Norwegian isolationism and backwardness.

Such cultural norms of attention make for a disjuncture between what one is required to think about global warming and what is normal to think about within Bygdaby. Seeing weather patterns as potentially connected to global warming requires the ability to visualize these global events—doing so magnifies their perceived seriousness and failing to do so makes them seem less significant, less real. *Bygdabyingar* have a collective backward focus, whereas the consequences of climate change will be felt in the future. An emphasis on the local is created through the activities detailed in chapter 1, including focusing on local events and history, maintaining a strong boundary between insiders and outsiders, reading the local newspaper, struggling against the outside world, and maintaining *lokalepatriotisme*, which I discuss next.

Lokalpatriotisme and Dialect

I described in chapter 1 the strong sense of *lokalepatriotisme* in Bygdaby. This local pride is expressed in their fight against "centralization" (a common theme throughout rural Norway) and in local language politics. Within Bygdaby, *lokalpatriotisme* shapes the boundary of attention, translating into a great deal of local identification, pride, and focus. There is a palpable tension between a focus on daily life and tradition within this community and the pressures of the wider world (perceived as both Norway as a whole and other nations). Local patriotism is expressed through knowledge of local history, pride in local achievements, and the use of dialect. Using local dialect affirms a sense of community, as Lene described it:

Kari When you hear the Bygdaby dialect, you know, for me—I can't explain it, not just the feeling that goes with the dialect, but also how people use the language, I get the feeling that people enjoy using their own dialect and value that, that it has meaning.
Lene I think it feels good to talk with other *Bygdabyingar*; then you can speak fast and well, and everyone understands what you're saying. It's a kind of community.

Place-Names, History, and Landscape

Experiences of time and place are interconnected. In Bygdaby, a prominent part of the sense of place has to do with residents' awareness of the history of that valley, from the activities of early farmers to the bombing and occupation of the town by the Nazis during World War II. The past is found in the landscape; past memories are located in specific places. As Gunnar described to me,

The history, it lies in the land. . . . So that the names themselves, they can give you a whole lot of information about the society. Which farms are oldest, which farms are youngest. . . . And you can find—if you look into the place-names—not the farm names, but the place-names; you can find history connected to the names. You know, the name is such that it has history attached to it. So that all at once you are going back to the society that was—like if you go to the areas that were uncultivated, there is a specific place they call "Olabukset," and that place is connected to a pointed stone rock that stuck out over a river, and the terrain was much on the other side. And that name was connected to an anecdote about a person with the name "Ola"; therefore, it was named "Olabukset" because that's where he jumped from. And on the farms, the different parts of the farm, the different fields, the different haymaking places had their names. And on earlier maps that actually had each path, each tree might have a name, so that the function, so that this is a very fine network of reference points.

I was fascinated with the enormous number of place-names for rocks, slopes, and small depressions, the stories that went with them, and their long histories. I was especially impressed by the extent to which people know and continue to use these names today. With the presence of place-names, or *stedsnavn*, the history is literally stored in the land. Community residents are reminded of these events as they go through their lives. To notice a particular place and remember what happened there as you walk by it re-creates a ritual of sense of place and history. When community members do this together, they build and reinforce collective memory, which reaffirms not only a sense of the local, but the sense of order and ontological security described in chapter 1. As I walked with the local walking club, they spoke the names of the places we walked through and told stories of the things that had happened there. A passage from my field notes captures what I experienced on such walks:

We continue walking, they stop and talk about a place where a new park will be built by the river. They explain it to me. We have come around the corner now, in the section of peninsula between two rivers. The group has stopped ahead of us to look at the buildings of the new language institute. Other members are talking about words that are dropping out of the language. "Hefte?" or something like that, "that means cold," one woman says and turns to find someone who is listening, [but] she ends up looking at me. We continue walking. At the next building, one woman remarks, "Here is where we had guests for five years." I suspect that she is talking about the occupation during the war but wait to see what happens next. She continues, "Guests in green uniforms who were called *tyskere* [Germans]." Much of the talk in the group seems to be about the history of the different places that we go [by], the buildings, the land. It is a kind of talk that is making sense of things, putting them in their place, bringing information up to date, "this building was used for this, what are they doing here now" kind of talk.

As we walked, group members were saying the names of places, telling stories, making order in the world.

The Local Paper
Besides local pride, dialect, and place-names, an important site for the construction of time and space is the local paper. Everyone in Bygdaby subscribes to the local paper, the *Bygdaby Posten*. In fact, as Lene, the paper's owner told me, the paper literally has 100 percent circulation:

We don't get so easily out to the remote places around, but we are very curious, and we really want to follow what is happening, so we get the newspaper delivered [at] home. It has something to do with companionship, too, such that the local newspaper in Norway—it is doing very well. It is the bedrock of the newspapers. Everyone wants to have their local paper, wants to read about the neighbor, wants to read about everything that is happening, different areas, wants to read about deaths and births, right? And everything that is happening in the community—there are, you know, the announcements. And if you are involved with something, then you'd like to see it written up afterward.

As people browse through the front page, featured stories, and the sports and classified sections, the events of the local community are there in print, pronounced as meaningful. Even I made the front page of the paper one day, not for my research, but for skiing through powder snow at the right time and place at the local ski area. The next day I was there in full color, and folks all over town commented that they had seen me in the paper. All this gave a sense of the significance of immediate local events and directed collective attention toward the events in the surrounding valley.

Reading the *Bygdaby Posten* works to create a sense of the local not only by covering the local people's lives and the events in Bygdaby, but, even more important, by doing so in a way that fails to connect those local events to the larger world. In chapter 2, I described how newspaper coverage of the unusual local weather patterns in 2000–2001 did not discuss theories on what might be happening or make links to weather patterns in other parts of the world, even though England (again, closer in distance than northern Norway) experienced extreme flooding that same season. Local news in general did not make links to events happening in other parts of the country or world. In so doing, the local paper affirmed a sense of boundary, a distinction between local events and the rest of the world. This boundary created a sense of security, the importance of *their* place and *their* history for *Bygdabyingar*, that it was indeed unique, that there was a mentally manageable local reality set in contrast with the wider world of chaos and changing times. The desire to maintain a sense of ontological security, of place and tradition, of local identity—these are not bad things in and of themselves. Like Ann Marie, the woman who appreciated Bygdaby anew upon returning from abroad, I too loved the sense of rhythm and meaning in Bygdaby daily life, the sameness of all the houses, and the predictability and security offered by the connection of mundane activities of the present to the activities of the past. But it may simultaneously be important to understand the multiple consequences of this sense of reality.

Producing Socially Organized Denial: Emotion Management and the Double Reality

Norms of emotion, conversation, and attention did not seem to provide much opening for thinking or talking about climate change in everyday life in Bygdaby in 2000–2001. But few sociologists view culture as statically imposed on individuals who lack all agency. We know that behind the scenes that year the people in Bygdaby expressed considerable worries. How and why did these norms come to be? In Bygdaby, there were active, observable moments in this period that, although fleeting, pointed to the role of emotions and culture in the active generation of community members' nonparticipation in affairs involving global warming. In contrast to existing studies that focus on the public's lack of information or concern about global warming, the evidence presented in this chapter and the next shows that holding information at a distance is actually an active strategy for people as they negotiate their relationship with climate change. The notion of socially organized denial emphasizes that ignoring is done *in response to social circumstances* and *carried out through a process of social interaction* (Zerubavel 2006). That is to say, at this time people had a variety of methods available for normalizing or minimizing disturbing information. These methods can be called strategies of denial. It is significant that what I describe as "climate denial" felt to people in Bygdaby (and, indeed, to people around the world) like "everyday life." Nonresponse to climate change was *produced* through cultural practices of everyday life.

In the next section, I build on what we now know about norms of emotion, attention, and conversation in Bygdaby to describe how people produced this double reality at the beginning of the new millennium. This framework builds on three strands of important theory: Ann Swidler's conception of culture as a tool kit, Arlie Hochschild's framework of emotion management, and scholarship on the importance of conversation for a sociological imagination and the formation of political power (Eliasoph, Lukes, Gamson, Arendt). The view I share here also incorporates the aforementioned concepts from social psychology and presumes that all social action takes place on the groundwork of Norwegian political economy.

First, recall Ann Swidler's (1986) metaphor of chunks of culture as "tools in a tool kit" to describe the set of resources available to people in a given culture for solving problems. Imagine a group of people skiing in the mountains. It is February, but the snowpack is much lighter than usual. If on this trip a skier raises the topic of historic land uses rather than changing snow patterns, she is picking up an existing cultural tool. She is following a cultural norm of attention (a focus on a localized history, with boundaries of both time and space), that may well be supported by another cultural artifact, a physical trail marker listing the place-name, which is likely hundreds of years old. Here both the norms of attention and the physical artifacts are available tools. Imagine as well that the speaker is very concerned about climate change. So as Arlie Hochschild's work prompts us to do, we notice that in choosing this topic of conversation, the speaker is both following emotion norms and perhaps managing her own troubling concern about the lack of snow by changing her thoughts to focus on a topic that is both personally reassuring and considered normal for the social context.

Of course, at the same time as she has re-created a world of safety, one in which climate change does not appear to be relevant, she has also contributed to producing the double reality and shaped the conversation in such a way that political talk does not emerge. What we have unpacked in this exercise is one observable moment in which climate change might have emerged as a political topic, but did not. In fact, this story is not fiction, but a description of the events and conversations that happened repeatedly during my time in Bygdaby.

To explore further the significance of this one moment, we can imagine what might have happened had the skier chosen a different topic of conversation. Imagine as the group comes over a rise, she exclaims, "Wow, in all my years I have never seen so little snow here in February! We keep breaking records for warm temperature and lack of snowfall. This makes me really worried! Did you read about how the youth ski team had to take the train up to Nordalen for practice this year? That has never happened before. What will this do to our winter tourism?"

And one can imagine how if the conversation continues along these lines that over time (and perhaps in repeated conversations), community members can generate ideas about taking some action. They might start with contacting members of city council and articulating their concerns or writing letters to the newspaper editor. They might convene a session of the Saturday Academy on the topic of local climate impacts. They will begin to make visible the significance of climate impacts in the community. They can use these local impacts as a springboard to draw national attention—for example, pressuring the Norwegian government to take stronger action on national emissions targets or a stronger stance in international agreements. The conversation that started that day on the mountain would therefore have engendered the kind of meaning-making power that would have made Hannah Arendt proud.

Social norms in Bygdaby are a source of pressure to conform, as Eirik articulated in the chapter opening: "There were many things that we felt that we should have done, but we did like the others anyway, right. Because you have to fit into normal society. It is difficult. You can't just sign yourself out, right? If you did everything entirely ideal you would

Table 4.3
Conversational Tactics and Strategies of Emotion Management

Humor
Knowing the facts
Controlling exposure to information
Shifting attention elsewhere
    Finding a *haldepunkt* in local community, tradition, and past history
    Focusing on something you can do

---

be an outsider in the society, and then you wouldn't get anything done either." At the same time as social norms exert pressure, complying with them serves as a tool or tactic for emotion management. And referring to such norms as the justification for one's inadequate actions can in itself be seen as a denial strategy, a way of minimizing individual responsibility. Opotow and Weiss describe "diffusing responsibility" as a form of "denial of self involvement," in which people, "deny personal responsibility for an environmental harm by seeing it as the result of collective rather than individual decisions and actions" (2000, 481).

Using this tripartite strand of theory, we move now to examine a series of culturally available strategies that serve as tools to achieve selective attention and perspectival selectivity and thereby to manage thinking in such a way as to manage emotions. In the case of Bygdaby, emotion management, conversational tactics, and strategies of shifting attention have become cultural tools that people use to keep disturbing information about climate change at arm's length. I examine the use of humor, conversational control, controlled exposure to information, and strategies of shifting attention. In the examples that follow, disturbing emotions are normalized and managed to fit social expectations, and through the course of conversation people re-create social norms by drawing on cultural tools. Notice as well that through all normalized activities we see the creation of the double reality—the *process* through which climate change is kept out of the sphere of everyday life. These strategies are summarized in table 4.3.

Irony, Cynicism, and Humor as Emotion Management

At one party three of our neighbors, all farmers—Torgeir, Harald, and Martin— were discussing where several of their friends were on that particular Friday night. Several community members had been invited to the "Julebord" or Christmas dinner party of the local *miljøverk* (in charge of disposal of "waste"—i.e., garbage, recycling, and air pollution). Martin said he didn't know about the Julebord. "Oh," Torgeir teased, "You were supposed to be invited. Maybe if you stop burning fires, they'll invite you next year."

—Field notes, December 22, 2000

In chapter 1, I described *Bygdabyingar's* active practice of telling jokes. According to Ann Marie, "I think it's quite old in our tradition to make fun of each other." As an American, I found jokes in Norwegian were the hardest thing for me to understand. I would often miss the punch line, understanding the words but not why the story was funny. But it was clear to me that humor among the *Bygdabyingar* plays a significant role in all sorts of social interactions. They have jokes about human behaviors and nicknames and sometimes even tell long, involved humorous stories. Humor and joking serve multiple purposes and play out in many complex ways—keeping people who try to do things differently "in their place," creating a sense of community, the feeling that "you're one of us," or merely serving as a form of clever word play with nuance and double meanings. Hollander and Gordon describe humor as one of the important tools of social construction (2006). Seemingly playful interactions can carry weighty messages about what is expected from oneself and others. At parties I attended, mildly insulting jokes about how drunk someone was at a previous party or what silly thing a friend did last week were passed back and forth like hot potatoes.

This practice of telling jokes has no doubt many social functions and effects, but several features of joking and the use of irony stood out with respect to navigating the issue of climate change during the time I was in Bygdaby. First, humor was a way of enforcing cultural norms through teasing people. In this sense, humor was used to send the message that it was better to play it cool and maintain an image of "all is well." Just as 15-year-old Trudi told me that caring about the world's problems "wasn't cool," caring sincerely and deeply about an issue was bound to make one the target of ridicule and teasing attention. For example, Soren, the one active local "environmentalist" in town was teased when he was seen driving a car.

Second, jokes and cynicism can also be understood as reflecting issues that were taboo. Joking about serious or socially taboo topics was a way to relieve the tension associated with these issues. Although there was little political discussion about climate change, I observed several examples of jokes associated with it. Climate change even made its way into fashion and advertising that winter. I described in chapter 2 how the lack of snow became the subject of a playful bid to sell advertising in the local paper. In another example, the fashion spread in the monthly train magazine was entitled "The Greenhouse Effect." In it, models displayed the latest fashions from inside a glass building surrounded by leafy plants. In this context, climate change was not only normal, but also hip, funny, entertaining.

Third, joking served as a strategy for emotion management. Eliasoph found that among her most informed but still inactive group, "teasing let members keep the wider world at arm's length" (1998, 107). Jokes and humor could be used as a way of maintaining conversational control over topics that were troubling. The issue of climate change evoked feelings of powerlessness and fears regarding ontological security that contrasted deeply with the norm of being in control. Eliasoph noted that for one group she observed, "displaying control in person was a way of making real control over the problems seem irrelevant" (158). For Eliasoph's respondents, "Cynical solidarity relied on first evoking the world's problems to show that I recognize the problems and, along with you, am not a 'bubba'" [dumb, country redneck]. The second step was to say why the problems did not affect me" (161). The people she observed used humor as a way to say that they were "impervious" and "somehow exempt" from the problems they discussed. "The world is a problem, but they always ended by asserting that they were untouched, not implicated, in control after all, laughing" (162). Humor was used to create the sense that "the problem affects other people, not us" (162).

In this section's opening passage from my field notes at the party, Torgeir was teasing Martin about making too many fires. Fires were a problem not only because of the particulate matter they produced, but because of their carbon dioxide emissions, which in Bygdaby were strictly controlled. Here the issue of climate change was almost visible, but in the form of a joke. Jokes and humor were used to as a tool for emotion management through making light of serious situations.

Another incident I jotted down in my field notes confirms this conclusion: "During our conversation in his office, Arne described the thoughts that came to him about climate change: 'Everything could go badly. That's it for Norway. One is always concerned first with oneself. I think we understand that things can go badly for those in tropical regions. How about Norway? Well, maybe then we can grow oranges in Norway at least.'" This joking phrase drew on the widespread knowledge of what life in Norway is like, what the climate is, and what Norwegian farmers grow. Norway is, to use the phase in a title of a recent book, "et langt, kaldt land nesten uten mennesker" (a long, cold country nearly without people). Norway is *not* a place where one grows oranges. Probably many other changes would come to pass before people started "growing oranges in Norway." The suggestion is very flip, as though things will smoothly flow from one climate regime to another without major hardship or major changes in social structure and ecosystems.

Knowing the Facts
In a crisis situation, people long for helpful information. Ten years ago my father had a major heart attack. Our family gathered to be with him in the hospital. As we hovered in shock outside the intensive care unit and later alongside his hospital bed, we became fixated with information—the details of the surgical procedure used to stop the heart attack, specifics of his body chemistry, the chemical properties of the various drugs he was taking. We did not discuss the forbidden question that was on everyone's mind, "What if?" We did not discuss our fears of death, of the unknown, of losing him, although these thoughts showed on our pale faces. We did not break down. Instead, we calmly but urgently drilled doctors for information, as if knowing the details of blood chemistry or surgical procedures could save us from our pain and fear. In the face of the possibility of losing him, we became fixated on the only thing we could control: information.

Displaying control over information as a strategy for control appeared to be in active use in Bygdaby in the unusual winter of 2000–2001. Norway is a modern, technologically advanced society. *Bygdabyingar* are well educated and informed and have a certain level of pride with respect to their nation's leadership on international environmental and humanitarian issues. In this context, it is not surprising that there was a degree of focus on technical discussions that year. The mentioning of facts, figures, and data was used as a rhetorical device to construct a sense of control and for speakers to present themselves as being composed and smart. With this strategy, one could be clever by displaying control over technical information. The emphasis on knowing the facts reflects the cultural norm of maintaining control described earlier. At the same time, following the norm served as a strategy for emotion management. Again, the sense that "the problem affects other people, not us" is an example of what Opotow and Weiss call "denial of outcome severity" (2000, 481), as described earlier. Information is known and at the same time taken for granted. Life goes on.

I also noticed gender differences in the use of this strategy. I had the sense that whereas both women and men sought to control information, men appeared to use this strategy more often than women.

"We Can't Dig Ourselves Down into Depression, Right?" Controlling Exposure to Information
In chapter 3, I described how too much information evoked community members' feelings of uncertainty, and so the information was carefully managed. The strategy of controlling one's exposure to information reflected the emotion norm of optimism. Educators and activists themselves had to be careful not to become overwhelmed in order to continue their work. They, too, used control of information as a means to do this. Åse told me that in the meetings related to her work with Amnesty International people didn't talk about how bad things really are: "But when I read about people in these countries and such, I think it's really heavy information that I'm getting. I do as much as I can, and we can't focus on what's painful. We don't go in and have meetings and talk about how gruesome everything is. We talk about how it is, but we can't dig ourselves down into depression, right?"

Her tacked-on interrogative "right?" explicitly indicates that this need not to get too depressed is a taken-for-granted, shared norm. In another example of managing emotion and controlling exposure to information, Åse explicitly linked her actions in choosing not to read particular material to the norm of optimism and the need to keep from "thinking about that all the time."

> I think it's difficult to learn about so much that is so icky. And I read very little of the annual report that comes out once a year—and I am not able to read about everything that is in there, I am not able to take in all the details, I just can't. It is better not to know everything. So now when we have a case, you have to get into it, but you have to in a way concentrate on what you are doing and just cut out very much of what you know. When it comes to our group, we focus a lot on positive things—helping! When I am active, it's better that I write a letter than just giving money because I don't have time to work so much. And the cards that we sell, very colorful, nice cards—and it says on them about supporting the group. So you can do this here, without thinking to much about money, about plaintiffs, about killings and things like that. So we hint that we just need to do a little and it will help, and that way everyone can be involved.

Here Åse describes knowing about so much nasty stuff as if it is a heavy load and how she reads very little of the details—that it is in fact "better not to know everything."

Shifting Attention

Although one can control exposure to information by not attending public talks on climate change or reading the details of abuses in human rights reports, such information sometimes does come into our immediate view. What then? Here people can use the cognitive strategy of shifting attention. Recall how Lise commented, "There is a lot that is negative. Then I become like—yeah, pffff! . . . And so it's good that I don't allow myself to think so far ahead. But when someone has something that they are working on, in relation to what you are trying to influence—then it's, like, okay to be optimistic after all."

The most effective way to manage unpleasant emotions is to turn one's attention to something else. The strategies of turning one's attention away from the negative either by not thinking too far into the future, as Lise put it, or by focusing attention on something positive were common among *Bygdabyingar* in 2000–2001. Nilsen made similar observations in her work with Norwegian young people. She writes that "these interviews so far suggest that these young Norwegians are uncertain about what the long-term future well bring in their personal lives. This uncertainty is kept at bay by not looking too far ahead, either in time or in space" (1999, 181–182).

Focusing on Something You Can Do

Similar to the strategy of controlling exposure to unpleasant information is the strategy of focusing on something that one can do. In her interview, Liv described how in her human rights work information about events in other countries can make people feel vulnerable. She described the need to "turn information" so that people will see that something can be done:

Kari But how can you talk about a thing so that it isn't . . .

Liv They are insecure when they hear that children are in jail and stuff, so one has to turn it, present it so that they understand that there is something that we can do about this. What can we do? Yes, we write letters. You, too, can begin to write letters. You have to focus on something you can do.

Kari Yes, something you can do.

Liv Yes, otherwise they—we become very helpless. And it's like that also when I talk about that case, or I spoke a bit fast so that I didn't, it was a kind of "show night," so I couldn't focus too much on that there, and so I said what was wrong, and then I turned right away to "now you can be with us and write letters." And they understood that it is of use, with speeches and writing letters, and people tortured, you don't become so, you have at least helped a *little*; of course, one can't be involved in everything bad that is happening in the world, but one can do A LITTLE.

Here Liv describes speaking "a bit fast" about what is wrong and "turning right away" to being with others as strategies to minimize the severity of the situation that she is trying to educate people about. Emphasizing that writing letters would mean being *part of a group* worked against the feelings of hopelessness that can come from isolation. Furthermore, she emphasized that there are things that people can do. Why, then, were the teachers in Bygdaby so careful in teaching their students about climate change? Again, the need for optimism was linked back to unpleasant emotions, the underlying fear of powerlessness and hopelessness that the issue of climate change raised.

Finding a *Haldepunkt*

Traditions which appear or claim to be old are often quite recent in origin and sometimes invented. . . . [T]hey are responses to novel situations which take the form of reference to old situations. . . . "Invented traditions" have significant social and political functions, and would neither come into existence nor establish themselves if they could not acquire them. . . . Yet it also seems clear that the most successful examples of manipulation are those which exploit practices which clearly meet a felt—not necessarily a clearly understood—need among particular bodies of people.
—Eric Hobsbawm and Terence Ranger, *The Invention of Tradition*

Bygdabyingar have a remarkable sense of place, local identity, and collective history. I have illustrated how community members actively produce this sense. One might ask, Why? At one point in time, Bygdaby *was* isolated from the rest of the world. True, the people there have always traveled—across the mountains to other communities or to the coast for commerce—but in the past the extent of travel was significantly less. Fifty years ago people did not take vacations in Greece or the Canary Islands; they did not watch images of war-torn African nations on their television sets; and they did not import food, clothing, and labor from all over the world. But now isolation, local focus, and emphasis on tradition are being actively maintained through a multitude of cultural practices. Why?

Torbjørn, a man in his thirties who was originally from the community but had moved away, explained that the emphasis on tradition in

Bygdaby serves as a *haldepunkt* or anchoring point for people in the face of changing times. He pointed out the use of traditional images in advertising and many so-called traditions are in fact very new. I mentioned earlier the fact that even many of the designs of the *bunad*s (local traditional costumes) are recent, and at least some sections of the cobblestone streets of Bygdaby had been recently added because people liked them. Hobsbawm and Ranger describe "the use of ancient materials to construct invented traditions of a novel type for quite novel purposes" ([1983] 1992, 6). On the one hand, the disjuncture between scales of time and place in Bygdaby makes the problem of climate change seem distant. On the other hand, we can see the construction of this sense of time and place as a strategic response to the needs of the present, as a strategy for managing troubling emotions associated with changing times.

Emphasis on tradition is linked to identity in *Bygdabyingar*, to knowing where you come from. Hilde, a farmer from a long-standing family in the community, explained it to me.

Kari So this is maybe a dumb question in a way, but why do you think, why do you feel, that it's important to take care of culture?
Hilde Well, I don't know, but—it is like when you know that your forefathers had something, that they used it—I think it has something to do with trying to give the next generation a glimpse of how—yeah, it has something to do with identity. That you know a bit about these things that are from the community.

Fifty years ago there wasn't such an emphasis on the past. In the years following World War II, people enjoyed their improving living conditions and worked hard to rebuild their nation. Instead of an emphasis on the past, the notion of progress and the future was relevant. As a way to make sense of their struggles, *Bygdabyingar* looked to the future as the time when the nation would be modern and their lives would improve.

Bygdaby and Norway as a whole have more recently been in the midst of rapid social, political, and economic change. After World War II, the Nazis left, and economic conditions improved, but former concerns have been replaced with a new constellation of worries including threats from radiation and climate change. In response, community members are looking backward instead of forward. Arne explained this to me on a ski trip in early January. We passed an older log cabin (maybe 100 years old, it looked as if the boards had been cut from a homemade sawmill). I asked Arne if the cabin was still in use. "No," he replied. I commented on how well preserved it is. "Yes, we have begun to be good at taking care of things. We weren't so good at it before, before, say, 1960." "Why not?" I asked. "The country was poorer then. They were focused on being modern. Old things reminded them of when they were poor and life was hard. They wanted to forget."

Ketil, a man involved with a local culture institute, similarly described the reemergence of the importance of tradition for modern Norway.

Ketil It has gone in waves. In the period from 1950 forward, old things were old, and they weren't of any value. But from the end of the '60s we began to, you know, see the value in it. But then came the "Japanese times," and all the young people had interest in was money and being rich, and all the old stuff, that was kind of. . . . But in the past ten years there has been a lot of interest in taking care of the traditions that are here. And particularly here in the school community, which Bygdaby is, people have been more and more concerned with their own culture. And with strengthening identity through cultural artifacts.
Kari Why do you think the pendulum has come back from earning money and being modern?
Ketil We have to find other values in our society than being rich, having a good time, watching TV, and all that. And so we start looking to our local community for that.

Here Ketil explicitly described the importance of tradition as a sense of grounding in the face of ambivalence about present society. If traditions are indeed invented for the needs of the present generation, it would seem that in the face of uncertain futures and a confusing wider world, ideas of tradition and links to the past serve as an anchoring point in changing times and provide a sense of security that at least some people believe helps them to deal with the larger world. Note that for Kjell Magne the past may serve as a source of inspiration for the present society, a better direction.

Kari And why do you think that it's so important to take care of this, or is it just for increasing tourism?
Kjell Magne No, I think that taking care of tradition, culture, roots—that gives a sense of security so that one learns to feel safe with oneself, and that makes it easier to go to other cultures and relate to other nations.

Several community members commented to me that it was the older generation in particular that was most concerned with the past and tradition. In looking backward, people found a sense of security that gave them stability and order in the face of the problems of modern society. I attended a meeting of the *Bondekvinnelag*, the Farm Women's Association. Nearly all the members of this group were in their sixties and seventies. There was also an *Ungdomslag*, a young peoples' group of the Farmer's Association, but it had less than half the membership. Lars, a friend of ours, learned that I had been attending the Bygdaby *Utferdslag* ski trips, and he told me, "It's mostly those folks who spent their lives in offices who now go on those outdoor trips in the mountains. When they were younger, they took professional jobs, sometimes moved to the city. Now they are middle-aged, and they wonder if they made a mistake; they are looking to get back to nature."

Looking backward, paying attention to the local—each is an example of selective attention. Selective attention can happen indirectly by selective exposure (what news sources people use, what people they spend time with, what email lists or social sites they subscribe to). To some extent, people are capable of making choices about what enters their awareness through how they spend their time, what they read, and who they interact with.

Socially Organized Denial: Ideology, Emotion, and Power

Several sociologists have isolated explicit strategies of action or "tools" that individuals use to construct reality.[1] These authors tell us that social norms of attention, conversation, and emotion—that is, the social standard of what is "normal" to think and talk about and feel—are powerful, albeit largely invisible social forces that shape what we actually *do* think and talk about and feel. Events occupy our imagination, our conversations, and our hearts, producing the sense of what is near and far, significant and insignificant, personally relevant or personally irrelevant. Everyday reality is structured through social, political, and economic institutions and produced through ordinary actions and practices, in particular following (and thereby reproducing) the interconnected cultural norms of what to *pay attention to*, *feel*, and *talk about*. Just as social norms of attention, conversation, and emotion create the sense of what is *real*, they also work to produce the sense of what is *not* real, what is excluded from the immediate experience of normal reality. In this chapter, we have observed examples of how in Bygdaby in 2000–2001 people managed the unpleasant emotions associated with climate change by avoiding thinking about them, by shifting attention to positive self-representations, and—especially in terms of the emotion of guilt—by framing disturbing facts in ways that minimized their potency. These tools were used to re-create a sense of an ordered, safe, and "innocent" reality. As I show in chapter 6, people in the United States describe the use of parallel tools in their own particular context.

We can see here an intimate illustration of what Steven Lukes (1974) calls the "third dimension of power." Although power in the form of outright coercion is a serious matter, it is also more easily recognized, identified, and condemned, especially in transparent and democratic societies. In contrast, this third dimension of power, which plays out in cultural denial, is so effective precisely *because* it is invisible and thus feels "natural." As Cohen notes, "Without being told what to think about (or what not to think about), and without being punished for 'knowing' the wrong things, societies arrive at unwritten agreements about what can be publicly remembered and acknowledged" (2001, 10–11). At the same time as norms of attention feel "just like everyday life," they reflect a particularly insidious form of social control. About the "third dimension of power," Luke writes, "Is it not the supreme and most insidious exercise of power to prevent people, to whatever degree, from having grievances by shaping their perceptions, cognitions and preferences in such a way that they accept their role in the existing order of things, either because they can see or imagine no alternative to it, or because they see it as natural and unchangeable" (1974, 24). This third dimension of power works in tandem with the first and second dimensions of power: outright coercion and the ability to set the public agenda. In the case of Bygdaby, we have followed how the control of what people perceive as normal is shaped via the social norms of attention, conversation, and emotion.

The notion that power operates in this realm of culture is also at the heart of Antonio Gramsci's (1971) concept of hegemony. Gramsci describes how power is maintained by dominant groups in society not so much through the use of overt force, but through securing the larger community's consent. Eliasoph points out, "The way people make sense of everyday experiences usually discourages them from thinking thoughts that might challenge the status quo" (1998, 232). Thus, as Nina Eliasoph puts it, "hegemony gerrymanders the boundary of perception," and people come to hold "a mistaken interpretation of how the world actually is" (Eliasoph 1998). What both Gramsci and Eliasoph want us to understand is that existing social relations are reproduced largely because people unconsciously consent to a smaller view of possible ideas (and I would add emotions). Gramsci developed the concepts of hegemony and ideology in relation to how top-down power is enforced in the context of Italian fascism. Although a handful of individuals in Norway obviously hold significantly more power in directing Norwegian petroleum production and economic activity, everyday community members are themselves complicit in the social organization of denial, as Gramsci shows us.

In connecting emotions with culture, performance, and political economy, this approach clarifies why it makes sense to conceive of denial as socially organized rather than as merely an individual phenomenon. This perspective moves beyond two limitations to present studies of how people view climate change: the focus on the assumption that information is the limiting factor and the focus on the study of individuals. Instead, I describe individuals *resisting* available information and doing so both *because of* and *through* social norms and interactions.

The management of information about climate change in daily life is part of both the management of emotions and the production of collective reality. Most research to date has examined denial on the level of individual psychology (e.g., defense mechanism) or political economy (e.g., corporate control of media). Yet in between the outright coercion of a corrupt government and the innate tendencies of our psychological defense mechanisms lies the realm of social action of everyday citizens. Here too social structure is encountered and reproduced. What individuals choose to pay attention to or ignore may have psychological elements but must ultimately be understood within the context of both the shaping of interpersonal interaction through social norms and the broader political economic context.

It is precisely because society organizes patterns of perception, memory, and organizational aspects of thinking that we must use psychology and sociology together to understand climate denial. To the extent that emotions, beliefs, identities, and cultures of talk are themselves socially organized, and to the extent that community members use existing cultural scripts as strategies to distance themselves from information, climate denial is a social rather than just individual production.

In addition to providing what I hope is a much richer explanation for public silence in relation to climate change, this chapter lays out an altogether new model of denial. Building heavily on Zerubavel's crucial observation that denial is socially organized, my work adds both emotions as a key *motivating factor* and emotion management as a central *mechanism* in the production of denial.

In the next chapter, I describe how individual acts of turning away from or avoiding thinking about the consequences or implications of information on climate change are also supported by the use of existing cultural narratives. In 2000–2001, when it came to the strategy of framing and of shifting attention to positive self-representations, Bygdaby community members had available a set of "stock" social narratives upon which to draw, many of them generated by the national government and conveyed to the public through the media. The source of these narratives is yet another important reason why it makes sense to talk about denial of climate change as socially organized.

5
The Cultural Tool Kit, Part Two: Telling Stories of Mythic Nations

> Norwegian adoration of nature is a vital ingredient in the country's national identity. Over half of the population has ready access to a cabin, the schools arrange annual obligatory ski days, and most postcards produced by the tourist industry depict nature scenes rather than cultural attractions.
>
> —Thomas Hylland Eriksen, "Norwegians and Nature"

While waiting for the train one afternoon, I spent a few minutes browsing through the postcard rack at the station. This postcard rack was located in one of the areas of town most frequented by tourists (most of whom are international). The rack displayed the images that Bygdaby has offered of itself for sale to others. There were postcards with dramatic images of snow-covered mountains, deep fjords, or Bygdaby from a distance so that one can see the surrounding farms and mountains.

Two images struck me that day. One postcard showed an old wooden pair of skis placed on the snow at the forest edge as though the viewer might step into them and be off for a ski trip. I was struck not only by the typical nature of this scene, the way it spoke to *Bygdabyingar's* love of skiing, but also by the timeless or premodern quality of the image of skis and nature. The skis were of the style found in our landlord's barn, in use up until the 1950s, but wooden skis have been used in Bygdaby for thousands of years, so the image might have been from any point during that time. The second card had an image of a child of four or five on skis, dressed in national costume and holding a flag. The card was so stereotypical that a person from Bygdaby would most likely purchase it only to send it either to a foreign friend or to other Norwegians as a joke.

Norwegian sociologist Thomas Hylland Eriksen (1993) has noted in his discussions of Norwegian self-presentation and national identity that postcards in Norway contain neither great works of art nor images of large cities, but pictures of a pristine and dramatic nature. He also writes that images of national identity are produced largely for the benefit of others, and these images are for sale to tourists.

I bought the two cards and caught my train. Later, as I watched the landscape from my window, I realized that the postcard with the child on skis was even more highly symbolic than I had first noticed. Eriksen describes five elements of national identity that are highlighted in public discourse: egalitarian individualism, honesty and sincerity, a connection with rural life, a relationship to nature, and "unsophisticated" practicality. Indeed, the card in my hand managed to combine nearly all the features of "official national identity" described by Eriksen. The presence of the flag emphasized the *Norwegianness* of the scene and the nationalistic nature of the image. The national costume, or *bunad*, that the child wore represented connection to a rural past (although some cities have their own typical *bunad*s, most *bunad*s come from rural regions). The child was skiing on a pair of wooden skis. The old skis also indicated a connection with rural life and with the past, but the fact that the child was outside on skis itself represents one of the most important and multidimensional elements of national identity: connection to

Figure 5.1
Postcard showing Norwegian child on skis. *Photo:* Normanns Kunstforlag.

Figure 5.2
Poster in store window in Bygdaby. The text reads "Norway, the cradle of skiing."

nature. The use of old wooden skis pointed not only to a connection to the past and tradition, but to the typically Norwegian values of simplicity and humility. Finally, the fact that the person in the photo was a child reinforced both a sense of innocence and purity and essentialist notions of Norwegians as *naturally* close to nature.

The image made me think of another image in Bygdaby, a poster in the store window of a frame shop on Main Street. The poster, which is from the 1950s, is a painting of a child in a cradle with skis coming out of the side of the cradle. The text over the image reads "Norge Skisportens Vugge," which translates as "Norway: The Cradle of Skiing." The use of the cradle in this image refers not only to Norway as the historical origin of skis (along with Finland), but to the notion that, as the expression goes, "Norwegians are born with skis on their feet and a backpack on their back."

As noted in chapter 1, stereotypes portray Norwegians as a simple, humble, nature-loving people who are concerned with equality and human rights. Residents of Bygdaby frequently emphasize the importance of simplicity to their lives, and they certainly are avid outdoors people. In the typical account, Norwegians have an enlightened sensibility, keep up with modern trends, yet live close to the land and keep one foot in a traditional past. They love to hike and ski and to spend holidays in the mountains. Their air and water are pure, and they enact strong environmental measures to keep them that way. An oversimplified and stereotypical view? If so, why, then, does it persist? Why is this official image so vigorously discussed and promoted? Is it meant merely to drum up tourist income for a brief summer season?

In addition to the role of norms of emotion, attention, and conversation discussed in chapter 4, the ways that community members understand and respond to global warming is organized by the more visible process of telling stories about their community and about Norway as a whole. This chapter illustrates how social institutions are key to the social organization of denial by describing the production and use of legitimating background narratives.

Various narratives, most either produced or reinforced by the national government and echoed by citizens, serve to legitimate and normalize Norwegian climate and petroleum policy. In addition, members of Bygdaby have a set of "stock stories" or social discourses about who they are and how the world is that they can use to deflect threatening information and uphold traditional versions of reality, thus maintaining moral order and the Norwegian sensibility. On the one hand, narratives of national identity frame background and context. By portraying Norwegians as close to nature, egalitarian, simple, and humble, these narratives of national identity counter the criticism that Norwegians face with regard to climate change and petroleum policies. These narratives of national identity assert a sense of who and what Norwegians really are that official government images, advertisers, and residents draw on. In this chapter, I argue that these discourses, which may seem trivial or unconnected to climate change, are in fact central to the process of socially organized denial. On the other hand, other narratives are more specifically linked to particular, contested climate and petroleum activities such as the expansion of oil and gas production or carbon-trading strategies. These narratives are produced by government ministers and conveyed via media to the public at large, where they are picked up and used in conversation. These narratives can be classified as forms of interpretive denial, perspectival selectivity, and claims to virtue.

One critical type of story is the story of *who a people are*, which is communicated through images of national identity. National identity has been the subject of increasing attention in Norway over the past decades, despite the fact that Norwegians themselves are very often conscious of its oversimplified and constructed qualities. What are Norwegians really like, and in what ways are they different from other people? Norwegians have an image of "a standard Norwegian" in their minds. This image is contested, but often in ways that in actuality reaffirm the original concept.[1] Eriksen writes that discussions of national identity, "never fail to arouse great passion" and that "in the early 1990s these issues . . . flared up with almost unprecedented intensity" (1993, 11–12). Drawing on ethnographic material from chapter 1, this chapter begins with a discussion of community members' relationships with nature and national identity. I describe not only how elements of national identity work to maintain distinctiveness (as emphasized in existing literature [see, e.g., Eriksen 1993]), but also how each individual quality represents (through connection to nature, equality and humanitarianism, connection to rural life and simplicity), engages, and deflects specific tensions associated with climate change and modern life.

I call the official Norwegian *selvbildet* (self-image) or national identity—the standard version of who Norwegians are and what their lives are like—"Mythic Norway." Idealized characterizations of Norway and Norwegians may be "true" to an extent, yet at the same time public images of national identity are a social construction—that is, they are hardly the only stories we can tell about Norway and Norwegians. In Norway, long-standing cultural values of environmentalism, simplicity, and equality are increasingly in contradiction to political economic relations (see, e.g., Sydnes 1996; Hellevik 2002; Hovden and Lindseth 2002, 2004; Rund 2002; Lafferty, Knudsen, and Larsen 2007).

Interpretive Denial

Interpretive denial involves the use of stock stories to frame potentially disturbing information about climate change in a more positive light. Social psychologist Morris Rosenberg notes that to the extent that people are able, "[they] tend to assign those meanings to events that will produce the desired emotions" (1991, 135). He describes this process as *selective interpretation*. For example, members of Bygdaby have a set of "stock stories" about who they are, or narratives of Mythic Norway. As noted earlier, Eriksen describes five elements of national identity that are highlighted, or in Erving Goffman's terms "overcommunicated" (1959), in public discourse: egalitarian individualism, honesty and sincerity, a connection with rural life, relationship to nature, and "unsophisticated," farm-based practicality. By portraying Norwegians as close to nature, egalitarian, simple, and humble, these narratives of national identity serve to counter the doubt and criticism Norwegians face with regards to climate and petroleum policies. Idealized portrayals of Norwegian national identity tell a particular story about *who Norwegians really are* that deflects attention from the fact that Norwegian wealth, political economy, and way of life are intimately connected to the problem of global warming—not only through individual actions such as automobile usage, but through the political economic structure that has created Norwegian wealth through the production and marketing of North Sea oil. Notions of Mythic Norway are portrayed in official government images and drawn on by advertisers and everyday people in Bygdaby.

Bygdabyingar also normalize information about global warming, using what Rosenberg calls "perspectival selectivity," which "refers to the angle of vision that one brings to bear on certain events" (1991, 134). For example, people may manage unpleasant emotions by searching for and repeatedly telling stories of others who are worse off than they are. Three narratives in this category—"Amerika as a Tension Point," "We Have Suffered," and "Norway Is a Little Land"—serve to minimize Norwegian responsibility for the problem of global warming by pointing to the larger U.S. carbon dioxide emissions, stressing that Norway has been a relatively poor nation until quite recently, and emphasizing the nation's small population size. For example, multiple newspaper articles in the national papers in the winter and spring of 2001 visibly listed the figure that the United States emits 25 percent of total global greenhouse gas emissions but accounts for only 4 percent of the global population. Although the United States must obviously be held accountable for its emissions, framing the figure in terms of total emissions and population makes the difference between the United States and "little Norway" appear very large. Per capita emissions in each country show, however, that the contrasts are not so large. In these articles, perspectival selectivity was used to create Opotow and Weiss's (2000) denial of self-involvement.

Robert J. Lifton coined the phrase "claim to virtue" to describe how the Nazi doctors in concentration camps who gave Jews lethal injections interpreted their genocidal actions in terms of compassion. From the doctors' perspective, their acts were compassionate because by killing people who were ill (or who might become ill), they were able to prevent the spread of disease in the camps. Through the claim that unjust acts are actually working toward an end that is opposite to what appears to be the case (i.e., saving the Jews rather than killing them), these acts are made acceptable. Two such claims to virtue have been used in Bygdaby and Norway to justify the expansion of oil and natural gas production.

The Elements of a Public Norwegian Identity: Modern Contradictions of Old Stereotypes

Romantic and utopian images of Norway are easy to come by: images of the land, the culture, and the relationship between them. The pages of travel guides and magazines, museum displays, and the official Web pages of the Norwegian government describe Norway as a spectacular land of mountains and fjords, of winter darkness, and summer midnight sun.

The community of Bygdaby might visually have stepped right out of these guides. A great example of natural beauty, it is nestled in a valley amid steep mountains, with the town beside a lake. It forms a picturesque example of a rural town embedded in nature. Surrounding the town are fields and pastures, and beyond them are the mountains where community members ski during the winter and take their sheep to pasture in the summer.

However, as much as some of the stereotypes fit, the image of Mythic Norway exists in sharp contrast to many everyday features of social life in Bygdaby. *Bygdabyingar* believe in equality but live their "simple," "equal" lives in extreme wealth compared with most of the world— including their immediate neighbors in poverty-stricken Russia. Despite their self-portrayals as simple, humble, and backward, Norwegians are on the whole a very wealthy people who, in addition to leading the world in per capita donations of humanitarian aid, work in a sophisticated and savvy manner to maintain their global economic and political status. Norwegians themselves are aware that their situation is not so straightforward as the stereotypes portray.

The first time I met Soren, we drove together to an environmental meeting in the next town. In the car, Soren told me, "There is a dark side to Norway. We seem to be the green country, but we are not really; we are very greedy. We have had more and more money, especially since the 1960s and 1970s. Rapid wealth, and it has done something to us. Look at this now, the FrP [right-wing Progress Party] is in the majority here. That is what the polls say. Politically I am more socialist. But the Nazis in the South, they look to Norway and say, 'There is a ripe ground for us.'" I ask if there are any Nazis here in Bygdaby. "No," he says. "But they are strong in Norway. And now that we are getting richer, we are giving less; it should be the opposite. Norway is exploiting other countries." "How?" I ask. "Now we want beef from Botswana because Norwegian beef is so expensive. We shouldn't have beef from Botswana until the people there are able to eat beef. And the brain drain. We are taking nurses from the Philippines. But what is the ratio of children to nurses there versus what it is in Norway? It is not right that we should take them away." I say that I think my own country can be very exploitative in a more explicit way, by, for example, going to war to control resources elsewhere. "Yes," he says, "but Norway stands on the side cheering."

Along the lines suggested by Soren, the most environmentally active person in Bygdaby, Norwegian oil companies are making a great deal of profit on cheap third-world labor as they extract oil from poor nations; Norwegians import resources from poverty-stricken nations, such as doctors and nurses from the Philippines and meat from various African nations. These activities provide cheap goods and services to Norwegian citizens but extract sorely needed resources from already impoverished regions. Despite their wealth amid a sea of poverty and political strife, Norwegians have also closed their borders to regular immigration since 1975. Norwegians claim to value equality, yet people of color report significant difficulties integrating into Norwegian communities, and racial tensions and instances of racism within Norway are both widespread and rising. A recent study by a Norwegian environmental organization compares Norway to other industrialized nations and ranks Norway near the bottom of the list on the majority of environmental indicators (Hille 1995). Geir, the environmental activist from Oslo, described the results of the report to me:

When it comes to the situation in Norway, I first have to say that with respect to ecological problems the situation is especially bad in Norway. There are few countries who damage the earth more per inhabitant than us, and furthermore we do poorly on many environmental indicators. We completed a report that compared the four northern countries and the six largest industrial nations with thirty different indicators of global environmental problems. And we did it again just a few years ago, and now Norway had sunk like a stone and was number 9 out of 10; only the United States was worse. And the cause was our richness. We fly a lot, we use a lot of cars, we are heavily dependent on industry. And if you look at classic environmental protection, we are not good because here in Norway we have the feeling that there is so much nature that we don't need to take care of it. And we are cutting most of the forests unecologically; we have protected one percent.

In short, when it comes to the issue of climate change, there is no shortage of contradictions between Mythic Norway and the realities of Norway's modern environmental policies and behaviors (see also Lafferty and Meadowcroft 2000; Sydnes 1996; Hellevik 2002; Hovden and Lindseth 2002, 2004; Rund 2002; Lafferty, Knudsen and Larsen 2007).

Normalizing Narratives

A visible, culturally prominent public consciousness of national identity serves at least four purposes that position and provide context for Bygdaby community members' responses to global warming. If information on climate change is a threat to people's sense of security, as indicated in chapter 3, how do people respond to such information? First, discourses of national identity reinforce a notion of Norwegian essentialism and exceptionalism—that Norwegians are naturally exempt from charges brought against others because theirs is a country with a small population, because they believe in equality and humanitarianism, because they are pure and close to nature, and so forth. Second, discourses of national identity mark a boundary between insiders and outsiders that is a key part of such essentialism. Third, stories of national identity connect to the past, to tradition, and to a sense of continuity of time and culture, thereby reaffirming a sense of stability of the traditional order and worldview, the sense that all is well. Finally, idealized portrayals of Norwegian national identity tell a particular story about *who Norwegians are* that in fact obscures actual lifestyle, political, and economic choices that contribute to global warming.

Here I describe public images of national identity and Mythic Norway as stories that serve to protect and legitimate particular versions of reality in the face of perceived threats. But I do not mean to imply that these accounts are necessarily "false"; rather, I wish to address why it is that Norwegians have put so much emphasis on telling this story. I am not saying that Norwegians should not be proud of who they are, but the fact that they need to present themselves so proudly in a certain light tells us something about the reasons why national identity is a focus in wealthy nations such as Norway.

Using Swidler's tool kit metaphor again, we can think of the elements of national identity as tools that people use to make meaning out of problems, to approach any conflicts between desired (mythic) notions of who they are and the realities of global inequity and environmental decline, to help them legitimate their actions and sense of place in the world, and to reinstate a threatened moral order. I believe that public interest in national identity is so strong in Norway exactly because the identity provides people with this set of tools. *Bygdabyingar* have available what I call "tools of order" and "tools of innocence." The former are images, ideas, and activities that are used to affirm a sense of how things are in the world—to create ontological stability—and the latter are ideas, images, and activities that are used to create distance from responsibility and to assert rightness or goodness. I begin my analysis with a discussion of four of Eriksen's aspects of Norwegian identity. We can also see how each of these representations engages and deflects specific tensions associated with climate change.

Connection to Nature

A Norwegian who lacks interest in nature and *friluftsliv* ("life in the open") may well be accused of being a poor specimen by his fellow citizens. . . . The wild and varied Norwegian scenery and clean environment comprise a source of pride to many of the country's citizens, and it may be the most important component in the standard image of Norway presented to foreigners. Instead of drawing on grand cultural traditions or a proud military history, Norwegian patriots (and surely visiting foreigners) may talk of their beautiful mountains, clean lakes and breathtaking fjords.

—Thomas Hylland Eriksen, "Being Norwegian"

The most interesting and complex aspect of Norwegian national identity in Bygdaby is the presentation of Norwegian connection to nature. The mountains and fjords of Norway have captivated the imagination of inhabitants, travelers, and artists for centuries. Skiing and hiking are widely considered to be typical Norwegian pastimes and significant ways of connecting with the land. The largest club in Bygdaby is the local chapter of the national hiking organization. In this community of 10,000 to 12,000 people, the hiking club had 930 members in 2000–

2001. When I asked *Bygdabyingar* what they liked best about Bygdaby, the most common answer was the mountains and wide-open nature:

My coworker Audun is sitting in the back room, her face red brown from a day of skiing. "It is so wonderful. I love it up there in the mountains so much. I could go every day." (In fact, she reports that she nearly does so.) "I don't need to be with people, it is so great I can go alone. That really is the best form of therapy. I don't understand that more people don't do it, there are so many people who could benefit from it. When you get up there with the mountains and the sky . . . and if you are alone, you know, you see the tracks of ptarmigan and hare, and maybe pretty soon you see one. It's so fantastic. When we got up to Norfjell today, there were cars everywhere, we arrived at eleven, just packed with cars along the sides of the road. But the mountains there are so big that it just absorbed them all. You see a few people over on that mountain there and over there, and maybe a few more in the valleys, but . . . they go in both directions from the road and that area is so big. . . . You get up there, and it is just mountains and mountains and mountains. We could see Storfjellbreen. It is so beautiful and fantastic up there that, you know, I really don't believe it. It's not possible to understand.

—Field notes, April 18, 2001

Ongoing connection to the land can be seen in the Norwegian notion of *fritluftslivet*, which translates as "the open-air life" and communicates the importance of outdoor and simple living to the culture. Norwegians express *fritluftslivet* through city-planning practices (80 percent of the nation's capital is park land), legal codes that give anyone the right to hike, camp on, or travel through private property (*allemannsferdselrett*), and extensive time spent out of doors: "Wherever they live, Norwegians have an exceptionally strong interest in outdoor recreation: 90 percent of the population gets out to the forests, mountains, or coast at least once a year, and the average person gets out more than sixty times a year. On any day of the week almost a fourth of the population spends about two hours doing some form or other of outdoor recreation" (Reed and Rothenberg 1993, 20).

Bygdabyingar have a great deal of pride in their mountainous land and their ability to survive in it. Not only do mountains geographically define Norway (because most of Sweden does not have mountains, whereas a large portion of Norway does), but there is also the sense for many that Norwegians are special because of the relationship they have with the mountains:

Norway's national identity gradually took the form of a lifestyle characterized by closeness to, respect for and love of nature, particularly the subarctic mountain landscape requiring great courage, strength and endurance from those who have to survive in it. Danes and Swedes were in this light refined and decadent city people, and the image of the thoroughly healthy, down-to-earth, nature-loving Norwegian was established as a national symbol. (Eriksen 1996)

Images of nature are prominent in national artistic works, and the national anthem refers to mountains and sea.

Hiking and skiing in Bygdaby are not only recreational activities, but practices loaded with cultural meanings that re-create a sense of moral order and ontological security. In this sense, ideas of nature and what is natural form part of the cultural tool kit that people use to make sense of moral dilemmas in their lives. Emphasis on a connection to nature plays a part of in-group/out-group constructions by way of highlighting distinctiveness. There is something very concrete about whether one knows how to handle cold winters, find one's way in the mountains, or slide down a hill on skis. The fact that Sam and I are skiers ourselves was an important piece of cultural capital in this community. We weren't from the town, but at least we skied. It made us less suspect.

Relationships to nature also serve as both tools of order and tools of innocence in Bygdaby. The act of hiking or skiing connect people to the wider nature, to a sense of human cultural past and continuity (because one often travels through areas that were formerly *seter*s, or mountain farms). In the face of the fears that community members have about climate change, the state of the world, and what the future might bring, being in the mountains provides a sense of reassurance that all is well. Emphasis on closeness to nature—whether in language, in images, or via the activities of going hiking and skiing—serves as a *haldepunkt*, an anchoring point in the face of uncertainty.

Going into the wilderness can be understood as participating in a national ritual. In this ritual, participants have set costumes and scripts. There is, for example, a standardized list of simple equipment (so standardized and simple that I could tell you exactly, down to the brand of chocolate bar, what you should take with you). On a trip to the mountains, one should wear older clothing, preferably a red parka, and so on.

This ritual re-creates a sense of Norwegians as close to and respectful of nature, as morally good. The ritual has power most importantly for its participants. For those involved, there is an experience of the continuity of time and order in the world, reassuring them of how the world really is. This ritual also works for the surrounding community, who watch their neighbors *ga pa tur* (go hiking) and who walk by houses with skis on the porch. Bystanders, too, get a sense of continuity as their neighbors carry out this tradition. This pattern of looking to nature as a *haldepunkt* is most practiced by the older generation, those in their fifties and older. In 2000–2001, the majority of members in the Bygdaby *utferdslag*, the local hiking and skiing club, was in this age group.[2] One friend of ours in his early thirties who spent a great deal of time skiing and hiking said to me, "I have never been a member of that organization. The people in these *utferdslag*s, the older people, they maybe could have lived the farm culture and didn't, and now they regret it. They are people who have worked to make the society what it is today, in business, in government. And now they are a bit nostalgic about what they could have had."

Emphasizing a connection to nature also serves as a tool of innocence. If one of the things that made it hard for community members to think about climate change in 2000–2001 was awareness of their own contribution to the problem (chapter 3), then to reaffirm, through rituals, that *Bygdabyingar* are close to nature is to imply that despite their rising materialism, petroleum development, and wealth, they too are pure, naturally good, even "natural" environmentalists. In the context of romantic notions of nature as pure, an association with nature provides a reassurance of original innocence.

Relationships to nature are a way to legitimate Norwegian "goodness" in the face of antienvironmental behavior for several reasons. In the modern context, romanticism, with its glorification of nature as a pure refuge from the ills of modern society, still holds considerable power. And nature on this view is the inert, unchangeable starting point or backdrop for culture. To say that something is "natural" is to normalize it, to consider it acceptable, and also often to imply that it cannot be changed. Thus, biological arguments have historically been some of the most powerful forms of legitimation of sex, racial, and class power relations. Norwegian sociologist Brit Berggreen describes how, "for many reasons, Norway has stuck to the self-image of the independent and innocent individual, almost grown out of Norwegian soil, rain and fresh air" (1993, 51). Relationships to the land have been described as the basis for a "native" environmentalism (Reed and Rothenberg 1993). Joar, a community member involved in the anti–European Union movement, used the Norwegian "love of nature" as a basis for this opposition: "Of course, we have so much nature. And you know that many Norwegians love nature and feel that it is clean and nice, and we like this idea of being environmentally friendly as opposed to black smoke and filth from factories."

Because of romantic associations between nature and purity, relationships to nature also serve as the ideal tool for navigating ambivalent cultural terrain (Sturgeon 1997). What better defense for a belief or stance than to be associated with nature? Many groups and movements have used relationships to nature in this manner. The wilderness movement in the United States has a strong romantic component in which nature is turned to as a source of purity in the face of a modern world tarnished by industrial activity and material greed. I have argued that association with nature has also been a source of legitimacy, voice, and thus political advantage for ecofeminists (Norgaard 1999). In another example of a community using a connection with nature to naturalize social relations, Michael Bell (1994) describes how English country folk use the idea of a "natural conscience" as a way to cope with the persistence of class relations they find unsettling. If you wish to define something as either pure or inevitable, a good starting place is to tie it to that which we do not recognize as having a history: nature.

In the Norwegian context, *friluftsliv* (outdoor activity) has an explicitly purist element. One often sees older "tried and true" equipment in use for hikes in the mountains, from simple backpacks to red parkas, wool pants, and even wooden skis (which are, interestingly enough, currently making a comeback in Norway). This is why there is, for many "skikkelig friluftsliv folk" (*real* outdoor types), an emphasis on simplicity in the woods. Cell phones and geographical positioning units are increasingly used in case of emergency, but the daily experience emphasizes old clothing and gear, although they must be in good condition. Nils Faarlund, one of Norway's best-known promoters of *friluftsliv*, refuses to wear "modern clothing" such as gortex or polypropylene. Although many consider his particular position "extreme," it comes out of a context in which simplicity is emphasized.

Bygdabyingar feel that spending time in the mountains literally makes one a better person. Being in the mountains is part of the ideal life for Norwegians, especially perhaps for Norwegian men. One friend told us, "My father, he had a list of things that a Norwegian was supposed to do, and on that list, absolutely, was to be a mountain man." Ketil Skogen writes, "It is the popular view that outdoor activities further a generally 'sound' or 'straight' lifestyle" (1993, 221). And Norwegian philosopher Gunnar Skirbekk writes that "to go skiing is not only healthy, it is good. If you go on a skiing trip through Norwegian nature, you are a good person" (1981, 21).

One day I mentioned Skirbekk's comments about how going into the mountains makes you a good person to a local man who is both religious and an avid outdoor person and asked if he agreed and whether he felt it was generally believed to be so. "Yes," he said, and listed a number of ways that a person is developed or improved through being in the mountains. He felt that being in the mountains challenged one to be a better person by developing humility (you can't be arrogant out there) and by teaching enjoyment of the moment, endurance, leadership, and the experience of being a part of the whole ("en del av helhet"). On my first ski trip with the hiking club, one man spent time describing the local hiking culture. He told me, "Vi er kanske det enest land som for god samvitighet ifra det" (We may be the only nation that gets a good conscience from it). The ritual of "going into nature" reassures bystanders of "what is true about Norwegians"—namely, that they are simple, hard working, and nature-loving people. It conveys a sense of their goodness. I do not mean to imply that *Bygdabyingar* engage in activities such as skiing or hiking *only* in order to purify their consciences! They do these things for many reasons: for exercise or stress release, to be with friends, for love of the outdoors. What I wish to highlight here is that among the well-known benefits that one can "get out of" participation in these activities, there are also these invisible benefits related to Norwegians' guilt about their role in climate change.

A Connection with Rural Life
Related to Norwegians' reputation for loving nature is their reputed connection to rural life. Images of traditional rural life are widely used in Norwegian advertising. The dairy cooperative that provides the majority of milk products consumed throughout Norway sells a kind of sour cream called *Seterromme*. Here the word *seter* is the name for the mountain farms that were the site of milk processing until the past generation. The use of the term *seter* makes reference to rural life and thus implies a sense of the past and of the purity of the product (because it comes from the mountains). The same dairy also sells *setersmor*, or *seter* butter. In Bygdaby, one can also buy Norsk Fjellbrod, Norwegian Mountain Bread, its packaging showing a picture of a skier holding an old-style ski pole in the mountains. The effectiveness of images of traditional rural life as a means to sell products affirms those images' cultural salience.

Connection with rural life is considered a significant value in terms of the proximity to nature and distance from the tarnish of urban areas. The conceptual creation of the Norwegian nation at the close of the eighteenth century occurred in large part when people developed a sense of Norwegian distinctiveness rooted in rural tradition. Scholars from the

Figure 5.3
Images of Mythic Norway in advertising: "Norwegian Mountain Bread" wrapper.

Figure 5.4
Mythic Norway in advertising: milk chocolate wrapper picturing cows in the mountains.

city traveled the land documenting folktales, language, music, dance, and clothing traditions from rural regions. As a result of this particular history, tradition in Norway is equated with rural life, and self-conscious concern with tradition is an aspect of connection to rural life. The Norwegian connection to rural life and tradition is reinforced by the fact that Norway remains relatively rural, by the use of images of rural life in advertising, by the presence of folk museums, and by the persistence of dialects in urban areas (to provide a small sample). Although the population of Norway is rapidly urbanizing—at a rate of 5 percent per year—the percentage of the population that remains rural is one of the highest in Europe. Connection with rural life as part of the Norwegian national identity (along with emphasis on equality between the city and the country mentioned earlier) has led to powerful support for government policies to protect jobs and living standards in rural communities. And the fact that about 25 percent of the population still lives in rural communities in turn reinforces the sense that rurality is what Norwegians are about.

Tradition as well as references to and connections with the past are hugely important for *Bygdabyingar*. Rural life is marked by numerous local traditions in food, building styles, music, dance, national costume, and especially speech patterns. Dialects in rural communities are highly localized. In many parts of the country, it is possible to tell where a person comes from within a distance of 20 kilometers (see chapter 1). Interest in tradition has been on the rise in the past 20 years, and emphasis on tradition and connection to the past serves as a *haldepunkt*, or anchoring point (see chapter 4). There is the familiar adage that "the second generation tries to remember what the first generation tries to forget." Images of rural life at the turn of the twentieth century are particularly significant because they tell a story of Norwegian distinctiveness. What role do these images play in modern society currently? Surely much of the focus on maintaining distinctiveness still holds. Yet Norwegian connections to rural life must also be understood in relation to the complexity and moral ambiguity of modern life. Proximity to nature and a simple rural life create a particular kind of moral order. Images of rural connection work to reinstate moral order by telling a story of simplicity, purity, and connection to the land.

Egalitarianism and Humanitarianism
In many ways, the connection to nature and rural life in stories of Mythic Norway implies an antimaterialistic, humanist national character.

Joar Yes, there is Jante's Law here in Norway—there was earlier. That you shouldn't be something.
Kari That you shouldn't be something?
Joar It was obnoxious to be rich. Yes, it was a little embarrassing to be rich. Yes, the idea of being successful, really successful, that wasn't seen as the ideal the way it is in the United States. The thing that was considered successful was to have a sense of community. That you could do something for the society. For everybody. A bit of a social democracy, you know? Labor unions. We stand together, you know? We are together, we are alike.

Joar described the declining (but to an outsider still substantial) influence of "Jante's Law," an often referred to social norm of not standing out. Equality is maintained by everyone doing the same thing as everyone else and thus not drawing attention to themselves. This norm is the very significant taboo against displaying wealth. Although it is true that times are changing, the "old" standards remain. Even Joar's casting of the change as negative is an example of enforcing the norm against the display of wealth.

Equality is a very relevant feature of life in Bygdaby. I sat one afternoon with Liv in her office at the middle school after hours. I wanted to speak with her about her work because it seemed that working with human rights issues was also a situation in which local people were exposed to their own privilege. I asked her whether she thought that Norwegians were particularly good at supporting humanitarian causes. In response, she described their tendency to give and share as a tradition:

Kari One thing that I have heard, but I don't, I don't even know where to find this information, but I have heard several times that Norwegians are very good at giving, with humanitarian . . .
Liv Yes. I don't know why, but in the old society here, it was like that with *dugnad*. Do you know what *dugnad* is?
Kari Yes.
Liv That people helped one another, and that was particularly true in the local society. You know they built houses and nurseries for children, and other things they did by *dugnad*. Someone was making a house, so they helped out. And then they got help the next time. So that is the *dugnad*'s role, to take care of one another. I think it must be something that sits, something we have in us from old. In the modern society, the new society, there isn't so much of it. And now we have these "collecting actions" like they have in the fall. Then of course we have those TV Actions that go to a good cause.

The linking of humanitarian aid to *dugnad* serves to naturalize a sense of Norwegian caring, making it "just something that we do, a part of us"—or, in her words, "something we have in us from old."

Egalitarianism and humanitarianism are linked for members of Bygdaby through the belief that all people should live a life of respect and dignity. Equality is a strong social value connected (along with the belief in social justice) to the moral imperative (Jonassen 1983) and the sense of moral order (chapter 1). Many scholars believe that equality has a long history in Norway, stemming from, among other things, the mountainous regions' limited ability to support large farms (Borchgrevink and Holter 1995). Without doubt, humanitarian concerns and equality between classes and sexes have been expressed as central aims of Norway's government policy since World War II. Despite current political trends, Norway retains a highly developed welfare state and a generally high quality of life for all citizens (i.e., good access to health care, high human rights, low poverty and unemployment rates). When compared with other nations, it has relatively little social stratification by race, sex, or class. Norwegians award the Nobel Peace Prize every

December; their Constitution Day parade consists of children rather than the military; and they lead the world in per capita contributions to humanitarian aid to developing countries. Norwegian leadership in international conflict resolution is reflected in the country's hosting of the Middle East peace talks that became known as Oslo Accords.

Equality is also a central part of the *allemannsferdselrett*, or right to freedom of travel and land use described in the previous section on nature. Several Bygdaby community members, one of them a 15-year-old, made a point of telling me about the *allemannsferdselrett*. On this occasion, Soren was driving me back from an environmental meeting held in a nearby community. We were engaged in a lively discussion about local customs and land use. He was very proud to tell me, "We have something that is very different from the way you do things in the United States. We don't have anything like your concept of private property. Here you can use the land, travel across it, camp there, fish."

I have already discussed the importance of equality as connected with the sense of moral imperative (chapter 1). The importance of egalitarianism and humanitarianism is reinforced through government policy, the educational system, publicly visible projects, land-use regulations, and general conversation—to give a wide-ranging but hardly exhaustive account. Belief in the equality between country and city dwellers has led to large efforts to keep the standard of living as high in the countryside as in the city. This value is of such importance that during the 1960s the national television station did not go on the air until television reception was available throughout the country. Government television and radio programming contains a humanitarian focus on the lives of people all over world. The school curriculum includes mandatory teaching on equal rights between the sexes (Vormeland 1993, 209). Oddvar Vormeland writes that the "educational system has emphasized freedom, equality and humanitarian values" (1993, 205).

A number of publicly visible projects can be understood as rituals through which *Bygdabyingar* and Norwegians in general collectively express humanitarian concerns. Every fall *TV Aksjon* (TV Action) is broadcast by the national television station to raise money for a different global humanitarian cause each year. On this day, volunteers go door to door to every home in Norway soliciting funds. Younger children participate in musical or other events in the street to draw attention to the cause. Regular television programming on the government station is suspended and replaced with programming that gives details on the issue and follows the fundraising effort. A second highly visible effort is *Oper-*

asjon Dagsverk (Operation Day's Work). All students at higher levels (approximately high school level) participate by engaging in a day's labor and sending the money they earn to a different humanitarian cause. Beyond these two major annual events are a host of smaller nationally visible efforts. On May 1, the Labor Party organizes a national fundraising event. Norwegians also participate eagerly in a program called SOS Barneby in which they make monetary donations or sponsor orphans in poor nations. The Norwegian state church has set up an organization called *Kirkens Nødhelp* (Church's Emergency Aid), which sends volunteers abroad for a variety of long-term projects. In addition, large international organizations such as the Red Cross are active in Bygdaby.

From my observations and interviews in my year in Bygdaby, I got the clear sense that although many *Bygdabyingar are* deeply concerned with equality, the notion that they are in fact egalitarian or humanitarian is a fragilely constructed project. Indeed, there is significant concern that Norwegians are not egalitarian or that they have become less so. As one community member told me, "I think we are becoming egoistic from our wealth, unfortunately. The young people who grow up today, they have way too much you know. They have everything material they could want. Nothing is a problem, we are getting spoiled in a way. There is too little motivation for them. For many." Many *Bygdabyingar* described being aware of these contradictions. As Liv stated in chapter 4, "We sit here so safe and good, even if there are some troubles here, but it's like a nursery school, the things we have here compared with other places. Just think of our neighboring nation, Russia, how they have it!"

In light of this uneasiness and the feelings of guilt described in chapter 3, emphasis on equality and participation in these social programs can be understood in part as a performance. Participation is sincere and caring, but it is also a "release valve," a way of coping with the contradictions of wealth and poverty. Norwegians *do* have a history of egalitarian and humanitarian behavior and ought to be proud of it. At the same time, the very self-conscious promotion of such information can be seen to serve other purposes. An emphasis on egalitarianism and humanitarianism as part of public identity serves as a tool of innocence, a way of minimizing threatening information about the nation's role in unequal economic and environmental relations. Norwegians are uneasy about differences in wealth, and these publicly visible humanitarian rituals provide them with a way of feeling better about themselves both individually and collectively. One woman I knew in Oslo said she and her six-year-old daughter went around to houses to ask for money for the fall *TV Aksjon*. She believed in the project. Yet she told me, "But it is kind of nasty, really, because it is also something we do to make ourselves feel better." Such activities can serve as strategies for maintaining a good conscience. As Liv put it, "And you understand that there is a point, with all this talk about people being tortured and the like, then you feel—you have at least helped a little. Of course, you can't get involved in all the bad things that are happening around in the world, but you can do a *little*. That's what I have done, anyway, and also you get a good conscience."

Knowing facts about Norwegian equality, history, and international contributions can also serve to ward off uneasy feelings. In December, I attended a very inspiring annual meeting of the local Labor Party. This meeting covered the activities of the party over the past year and discussed the current challenges. During the course of his talk, the local leader made a number of statements designed to promote pride in their achievement. "We have been the party who has thought about those other than ourselves. We are the ones who have created the welfare state. We have followed the slogan of taking from the rich and giving to the poor."

Although egalitarianism and humanitarianism are actual social values for which many members of Bygdaby strive, the self-representation of possessing these values is a tool of innocence that works to affirm publicly the goodness and rightness of Norwegian actions. In other words, Norway's national self-image does not include Norwegians driving somewhere to ski but shows them on skis already, out in the pristine forest. It does not show men working the North Sea oil rigs. And so on. The storytelling about Norwegians' political activism and their connections to nature hide another reality of the national economy.

Simplicity
The fifth element I wish to describe here is the sense of Norwegians as simple, humble, and a bit backward. Eriksen writes of the image of Norwegians as "unsophisticated but practically minded farmers" (1993, 22). The cover of the now classic book *Den Norske væremåten* (*The Norwegian Way of Being*) (Klausen 1984) depicts a person wearing a handmade sweater of a typical Norwegian pattern, a backpack, and the type of large red hat Norwegians call a *nisselue* pulled all the way down over his head. Another figure dressed in a suit, presumably an anthropologist, is pulling up the hat to peer at the person. The "typical Norwegian" is holding an oil can (not a usual part of the national

Figure 5.5
Book cover image of the "typical Norwegian."

image!) that is dripping oil onto the anthropologist's briefcase. In the background are mountains.

By now, it should be clear that the hat, sweater, backpack, dripping oil can, and mountains are stereotypes. The dripping oil can is meant to imply the stereotype of backwardness—the nation's oil wealth may be acknowledged in the image, but the use of gas in cars and other modern industrial technology that contributes to greenhouse gases is not. The sense of Norwegians as simplistic, unsophisticated farmers or as a bit backward is highlighted by the image of the *nisselue* (recall the phrase "pulling the *nisselue* down over one's eyes" as corresponding to the expression "hiding one's head in the sand").

A sense of Norwegian simplicity, humility, and straightforwardness is linked with the emphasis on equality and with the relationships to nature and rural life that I have discussed. It also arises from the widespread influence of the Lutheran Church in Norwegian society. This influence was probably greater a generation ago, yet Norwegian sociologist Marianne Gullestad believes the church's impact even today "may prove to be much more central to Norwegian culture than is commonly believed" (1992, 224). For *Bygdabyingar*, simplicity is emphasized in direct contrast to greed and materialism, which are frequently associated with the United States and condemned in official public discourse. The value of material simplicity is reflected in high federal taxes on "luxury" items such as cars, electronics, and cosmetics. Huts in Norwegian mountains should be simple in order to "signal an ideal of simplicity in lifestyle" (Eriksen 1993, 20). The simplicity of life in Norway is also reproduced on an experiential level through the sense of safety and the country's low population. The phrase "Norge er et lite land" (Norway is a small country) is a stock phrase that communicates a sense of community and lack of culpability. I discuss this phrase in more detail later in relation to the ways that Norwegians construct their own national identity in the act of condemning the United States.

The simplicity inherent in cultural homogeneity, although not an official aspect of national identity, clearly reinforces the experience of simplicity in daily life in Bygdaby. In chapter 1, I described how both tradition and government bureaucracy serve to create cultural homogeneity. Especially in comparison with the United States, a great deal of activity in Norway is organized by tradition, which simplifies daily rituals by reducing the range of choices available. The foods one eats (*skive*, bread slices, for breakfast, fish and potatoes for dinner), the way one behaves, the activities in which one engages on a Saturday (home with family) or Sunday (out for a hike), are marked by tradition. I was fascinated by the influence of tradition not only on building styles, but on the narrow range of paint colors used. I once asked my neighbor what she thought of the fact that almost all of the houses are red, white. or yellow. Interestingly, she misinterpreted my question as referring to the fact that some of the houses are *not* these colors. She told me that until the 1970s all houses had been either white or yellow, and barns were red. Then more colors such as brown and green began to be used on houses. That, she said, was OK, but other colors were pushing things too far.

Simplicity in Norwegian society is also reproduced through social policy. The Norwegian social democracy has bureaucratized and standardized many aspects of social life. Although bureaucratization creates complex rules, it also clearly lays out right and wrong ways to proceed, thus producing a strong sense of social order. The fact that life has become standardized has meant that things from the types of wines available at the government liquor store to the medicines in the government pharmacy to the words of civil marriage ceremonies are all exactly the same from the far rural north to inner-city Oslo. And again, until 1993, Norwegians had one national (government broadcast) television station so that people across the nation watched the same television news and entertainment programs at the same time.

The Uses of Norwegian Nationalism

Although images of simplicity of Norwegian life reflect certain realities, they also exist in contradiction to increasing materialism, which is in turn associated with global disparities in wealth and increasing environmental degradation. The sense that *Bygdabyingar* have of themselves as simple people does two important things. First, it asserts order in the midst of potential chaos. If life is simple, it is much more easily understood and controlled (although not part of Eriksen's description of the national identity, "maintaining control" is a culturally valued quality [Gullestad 1997; see also chapter 4]). Second, images of simplicity connote innocence.

Social scientists have paid much attention to the role of national identity in making a symbolic boundary and thus in maintaining the distinctiveness of a group of people (e.g., Andersen 1991). Yet if we view national identity as a construction, a story that is put forward, the particular content of that identity gives us a great deal of information about the cultural tensions that inspire such interest in national identity. Why have Norwegians chosen to emphasize particular qualities about themselves? Indeed, the specific qualities that Norwegians have chosen to highlight in their public identity—proximity to nature and rural life, egalitarianism and humanitarianism, and simplicity—serve as tools to reinstate a moral order threatened by the challenges of Norway's contemporary position in the world, including its role in large-scale environmental problems such as climate change. I believe that nationalism and national identity hold a great deal of cultural salience and force in Norway today because of their ability to naturalize and legitimize Norwegian sensibilities in the face of these contemporary challenges to traditional Norwegian values.

I have described these four elements of national identity as though they were distinct, but in fact they often intersect and reinforce one another. For example, perceptions of simplicity are reinforced through relationships with nature, which are in turn reinforced by images of rural life. Notions of nature and the natural are complex, multidimensional categories, and the sense of a connection to nature is by far the most complex social narrative. The latter idea, as I discussed earlier, plays off of romantic associations of nature with purity and innocence, thereby working to reinforce a sense of a native or natural environmentalism that works directly against the image of Norwegians as *miljøsvin*, environmental swine. In the face of global climate scenarios and Norway's increasing carbon dioxide emissions, the images of Norwegians as close to nature and as "natural environmentalists" provide a reassurance of "original innocence" and serve to legitimate Norway's actions and place in the world. Emphasis on egalitarianism and humanitarianism contradicts real information on the ways in which Norwegian wealth and high standard of living perpetuate and displace environmental problems onto others. Emphasis on simplicity locates Norwegians as "down to earth," humble, and modest and provides a contrast to the fact of increasing consumption. Finally, an emphasis on a rural connection, like the emphasis on relationships with nature, plays on romantic associations of nature with purity and innocence and brings to mind images of simplicity. Images of Norwegians as placing a high value on equality and humanity exist in tension with the presence of racism and Norwegians' position of great economic wealth. In the face of counterevidence, knowledge of how much Norwegians give in humanitarian aid (especially compared with other nations) or reminders of the numbers of people who participate in antiracism marches can be used as tools that reinstate moral order. Emphasis on an official image of Norwegians as concerned with equality directs attention toward positive aspects of national character. The focus on all these qualities can be seen as part of Goffman's "impression management" (1959) on a societal level.

"I Know Things Look Bad, but It's Not What You Think": Perspectival Selectivity and Claims to Virtue

For residents of Bygdaby, unpleasant emotions such as guilt and those associated with a "spoiled identity" can also be managed through more explicit forms of "spin." This "spin" is also part of the cognitive strategy of interpretive denial in which facts themselves are not denied, but are instead given a different interpretation. Stanley Cohen writes, "Officials do not claim that 'nothing happened' but what happened is not what you think it is, not what it looks like, not what you call it" (2001, 7). In Bygdaby, two such forms of spin were in use in 2000–2001: Rosenberg's perspectival selectivity and Lifton's claims to virtue.

Perspectival Selectivity

One style of social discourse in use in Bygdaby to normalize Norwegian relationships to the issue of global warming during my year there was perspectival selectivity. With perspectival selectivity, people may manage unpleasant emotions such as guilt by searching for and repeatedly telling stories of others whose behavior is worse than their own. The use of a narrative I call "Amerika as a tension point" is one such application of perspectival selectivity. The subtext of such stories conveys the messages that "we aren't so bad, look at them" and "you can't really expect us to do the right thing when others are so much worse." A second narrative, "Norway is a little land," deflects troubling information and their associated emotions with the subtext that "there are so few of us that it doesn't really matter what we do anyway." Both these narratives clearly hold "truth." When it comes to climate policy, the United States *is* much worse and more powerful than Norway. And given the size and complexity of the problem of global warming, what Norwegians do or don't do may be relatively insignificant. Although the plausibility of these tools adds to their effectiveness, it does not change the fact that they are used in a strategic manner for the negotiation of information on climate change and the management of emotions about it.

"Amerika" as a Tension Point

One afternoon in mid-November, a few weeks after our arrival in Bygdaby, Anne, Torstein, and their 15-year-old son pulled into our driveway and asked if we would like to join them on a trip up to their *støl*, or summer farm. Snows had fallen in the high country, daylight was decreasing day by day, and in the United States the nation was in an uproar over an undecided election. We drove in their car up steep and winding roads, following a route traveled by these farmers' forefathers between main farm and *støl* from long before the invention of cars. Anne asked us what we thought of the situation at home. "What a joke," she said, shaking her head, "America is such a joke."
—Field notes, November 18, 2000

I use the Norwegian spelling of the name "Amerika" to indicate that I am talking about Norwegian's view of the United States, what Steinar Bryn calls "Mythic America." Bryn has written extensively about Norway's relationship to the United States. His doctoral dissertation, "The

Americanization of Norwegian Culture" (1994), and other texts in Norwegian examine the historical relationship between Norway and the United States and consider why American pop culture has become so popular in Norway. He writes that Norwegians, like people in many other countries, have created a mythic image of the United States and that Norwegians are influenced by their image of Mythic America rather than by the actual United States: "Norwegians are fascinated by the myth, not the reality. That's why Norwegian tourists focus on Hollywood and Disneyland in the northern part of Los Angeles and not [on] the relations between Koreans and Blacks in South Central Los Angeles. . . . They are simply not particularly interested in the real United States" (1994, 52).

Although the Norwegian myth of America does not provide much insight into the United States, perhaps it provides insight into Norway (Bryn 1994, 54). "The powerful presence of 'America' in Norway in turn influences Norwegians' understanding of themselves. American values are often defined in comparison to or in opposition to Norwegian values, and America represents everything from openly expressed dreams to the most hidden fears. America becomes a Utopia or Dystopia from which Norwegians criticize or praise their own country" (Bryn 1994, 54).

Members of Bygdaby and Norwegians generally have an ambivalent relationship with the United States. On the one hand, "Amerika" has been a significant cultural model and source of inspiration for Norwegians. As previously mentioned, nearly one-third of the Norwegian population, including many people from Bygdaby, immigrated to the United States between 1850 and the early 1900s. For this reason, there are strong connections between Norway and the United States. The farmhouse in which we lived was built with money the farmer made while living for years in America. Other residents of Bygdaby also went to the United States and became successful. A statue downtown commemorates one such man. Those left behind received "Amerika *brev*," letters written to Norway from family members who had moved to the United States. The older generation of community members grew up with the excitement of receiving these letters and all their mysterious tales of life in America. The United States was also the ally who ended Nazi occupation of Norway and provided it with political protection during the Cold War when Soviet tanks patrolled the coastline. Yet alongside the fascination with the United States, there is a significant degree of resentment regarding its superior economic, military, and political power. Norwegians call themselves the "fifty-first state" at times, a somewhat cynical characterization that refers to Norway's close following of U.S. foreign policy and political philosophy. Images of America in Norway have changed over time. After World War II, America was viewed in a very positive light, but during the Vietnam War impressions of it became more negative. At present, community members seem to express both resentment of the United States for its dominant political and economic position and admiration of Americans for many cultural and political accomplishments.

Bygdabyingar know an amazing number of facts about the United States. These tidbits of information range from bizarre and trivial "minifacts" to significant aspects of political and economic history and policy. *Bygdabyingar* also have a number of stereotypical views of life in "Amerika." References to the United States served as a tension point in numerous conversations I participated in and overheard while in Bygdaby. These stories served as a way of saying, "We aren't so bad; at least we aren't like *them*." Stories about "Amerika" were told at strategic moments to deflect Norwegian responsibility and shortcomings and to support notions of Norwegian exceptionalism: "We may not be the best, but we aren't anything as bad as they are." As Bryn puts it, "Norwegians easily project certain aspects of their own culture onto America, thereby protecting their own cultural innocence and avoiding a deeper analysis of the Norwegian cultural identity" (1994, 4). There are many stereotypical images of the United States in Norway, but what is most interesting is not the images themselves, but how they are used.

One evening I attended a party down the hill from the farm where we lived. I met a young woman there who worked in a children's nursery. I asked her if it was true that there is a Norwegian law on the books that all schoolchildren must be outside for a certain length of time each day. "Yes," she said, it is so, and then she explained the law. "They wouldn't go for that in America, though," she spoke confidently and with a bit of disdain. "I was at a school once in Minnesota with friends, and it started to rain, and everybody panicked and had to get the kids inside right away. It was just a little drizzle! I couldn't believe it. In Norway, kids are outside year round." Given the widespread Norwegian belief that spending time out of doors makes one a better person, the subtext of this story is Norwegian superiority.

During community environmental meetings organized by members of an adjacent community and held in the smaller communities around Bygdaby, another interesting comparison was used. The meetings were basically an educational forum, part of a project to raise interest in the area's recently completed environmental plan and generally to raise awareness of environmental issues. During a slide presentation that touched on rising consumption, the speaker confidently cited a statistic that "70 percent of Americans value their cars more than their families." This rather absurd assertion was here used as a warning: "If we don't shape up, we'll end up like them." And implicit in such a message is the belief that "we are not like them. We may have our problems, but we are still doing better than *that*."

Chapter 2 opened by describing a conversation I had with one man in early January on a ski trip. When the topic of climate change came up, he said, "You know that they have been talking about climate change and that it is the fault of humans." I asked him what he thought of that argument. He gave a funny laugh, then said, "The United States has been reluctant to decrease their emissions." True. Yet Norway, like the United States, has also dramatically increased its carbon dioxide emissions. I also had the sense that mentioning the U.S. role in the matter served to manage emotions of powerlessness and guilt by giving the speaker something "clever" to say and someone to blame. Until this topic came up, we had been talking comfortably; now there was a sense of awkwardness. For the issue of climate change, there are no easy answers. The man's comment about the United States not only pointed a finger, but also fulfilled the need to have an intelligent response, for which there is strong local pressure (see chapter 4). Gaining control in the conversation about climate change is a way of managing the fact that one has so little control of actual climate change in daily life.

A final and particularly salient example involves the fact that many *Bygdabyingar* took the opportunity in talking to me to criticize U.S. president George Bush for his statement in March 2001 that he would pull the United States from the Kyoto Protocol on the grounds that it was "not in the U.S. economic interests." This comment was widely repeated and discussed in the Norwegian press and in public commentary and was mentioned in a local community speech given on May 1, which I quoted in chapter 2: "The Kyoto agreement is about cutting carbon dioxide emissions by 5 percent. And even that ridiculous pace was too much for the climate hooligan George W. Bush in the United States. The head of the USA's Environmental Protection Agency said that 'we have no interest in meeting the conditions of the agreement.' Well, that may be so. But it is other countries that will be hit the hardest from climate change."

Yet this is essentially the same move that the Norwegian government made in dropping national emissions targets, increasing oil development (which has been the largest source of increasing emissions), taking a leading role in the development of the carbon-trading schemes known as the Kyoto and Clean Development Mechanism, and shifting the focus from a national to an international agenda. In fact, a few weeks after Bush's statement, Prime Minister Jens Stoltenberg took a similar stance. In this context, criticizing the United States directs attention away from Norway's similar behavior and the negative emotions of a "spoiled identity" and sends the message that at home things are not that bad. The narrative of "Amerika as a tension point" is what Opotow and Weiss call "self-righteous comparisons," a form of denial of self-involvement in which the speaker casts himself or herself "as environmentally 'clean' and blameless in comparison to 'dirty' and reprehensible others" (2000, 481).

"We Have Suffered"

Three of us were sitting around a small table in Bygdaby's newest and most modern café. Lisbet and Åse were members of an international human rights organization in town and were actively writing letters on behalf of political prisoners around the world. Poverty and wealth are important issues for Norway not only because of the emphasis on egalitarianism, but because Norwegian wealth is very much a product of oil development. Our conversation so far had covered how they got involved in their organization, the kinds of work they did, and what it felt like. I asked them about the lack of poverty in Bygdaby and what it was like for them to know about the degree of poverty elsewhere in the world.

Åse Norway hasn't been rich so very long, you know. I have experienced the rise in wealth.
Lisbet Yeah, I know about going from hand to mouth.
Kari Oh, really?
Lisbet Oh yes. I began working in '62. And there wasn't much in the beginning, no. And when I took my education, I had almost nothing. But I did get a study loan. And it was good I had a job because I had to pay it back, you know? But I don't know, myself. I have a feeling that they are afraid to use up the rest of the oil money. Of course, we have a lot of money. But I don't have to look long at the television before I

understand how good we have it here. And I have traveled a bit, too, so I know how good things are here for us in terms of material things.

Åse But I think—I feel a bit bad for the Russians; I think it's completely gruesome what's happening there.

Lisbet There are many who are helping them, especially in the North.

Åse Of course, they are such a huge country.

Lisbet Yes, and how rich they could have been.

Åse Yes, and what has happened generation after generation. That things should be better, better in the future and so forth, and then . . .

Lisbet On the other hand, there are people in Norway also who don't have so much to play around with either today. They don't have much money today either.

Åse I think that there aren't so many who are suffering, not who really suffer. Not with lack of food. But it's like, it's like there isn't anything that is good enough for us anymore. Everyone wants things, and everyone wants a nice car, and everyone wants it right now. And if they don't have it, those who don't have it, then they are poor.

Here Åse and especially Lisbet used a number of conversational tactics to rationalize or normalize Norwegian wealth as compared with poverty in Russia. In addition to raising the issue that Norwegians suffered and were poor in the past, they mentioned that poverty still exists in Norway. This observation serves to minimize the degree of difference between Norway and Russia and thereby the sense that there is something wrong. They also mentioned that Russia is a large country that might have been rich, implying that "it's not our fault if they have messed up their opportunities."

The fact that residents of Bygdaby suffered or at least endured challenging political and economic circumstances in the past is certainly a reality. People in their forties and older have experienced the very rapid rise in wealth. Older generations clearly remember life during World War II, when the town was both bombed and occupied by Nazis, and many community members were active in the resistance. They remember the need to cover their windows at dark, the presence of soldiers in the town. A neighbor of mine, Bjarte, as a child of seven worked clearing the land for the airport. At that age, he witnessed the Nazis' shooting of a captive Russian soldier who did not work fast enough. Beyond the experience of occupation during the war, members of Bygdaby also experienced economic hardship. I met at least two community members now in their early fifties who grew up without flush toilets. One of these people, a woman, describes her childhood in the post–World War II years:

I was born in '46; we were five families living in one house, and I didn't have my own room. And it was in an apartment building of 25 families together. But we had a good life, we had food and clothes, we weren't poor, we were average. And my father, he left early in the morning, at seven, and was gone until five in the afternoon and worked also on Saturdays. He was home to eat lunch [*middag*] an hour during the day. And we didn't have a refrigerator, we didn't have a telephone, we didn't have a car. We didn't have any of the goods that we have today. But we had things very good. We were happy, and we played and came up with things and had a good life together with many friends. And if we got a bottle of soda when it was our birthday, that was a really big deal [*det var stor stass*, "lots of status"]; you were very happy for these little things.

The memory of having suffered can play out in terms of having paid one's dues, of rightfully having earned the comforts now available, and in terms of the view "We made it, so why can't you?" The latter view especially erases political economic relationships between Norway and the rest of the world. The subtext of such stories is that Norwegian wealth has been rightfully earned.

"Norge Er et Lite Land"

The phrase "Norway is a little land" is one I heard often throughout my time in Bygdaby, usually as expressions of Norwegian powerlessness in relation to the rest of Europe or the United States, but also at moments when Norway's relationship to world problems was in question. The phrase conveys the sense that no matter what Norwegians do, whether they do the right thing or the wrong thing, it is a "drop in the bucket" because there are only 4.5 million of them, which justifies Norwegian behavior or nonaction. Although the phrase often conveys a genuine sense of powerlessness, it also works to let people off the hook, creating the sense of "why bother."

Joar and I had been discussing his opposition to Norway's joining the European Union. When I asked him what role he thought Norway should have in the world, he began his reply by implying that there may not be much role for it to play because Norway is so small and thus almost meaningless.

Kari But what kind of a role do you think that Norway should take internationally?
Joar Internationally?
Kari Yes, in relation to other countries. What relationships would you think would be good for Norway to take?
Joar Well, we are of course a very small country, almost without meaning—if you think economically, we are completely uninteresting.

Kari But Norway has lots of oil compared with other countries.
Joar Yeah, yeah, OK. We are in fact almost at the level of Saudi Arabia. But it of course is an advantage that is meaningless. It doesn't really matter for us to argue.
Kari Without?
Joar Without them getting mad at us, because we are so meaningless. And in that connection, we are a bit, you know, peaceful, right? We have been involved in both the Middle East and . . . [here he refers to the Oslo Accords, and his second example is not spoken, just given as a gesture of the hand for emphasis].

Note that as the conversation continued, Joar used Norway's small economy to illustrate that the country's actions are meaningless. When I asked him about Norway's oil, he suddenly remembered the fact that Norway is, after Saudi Arabia, the second-largest oil exporter in the world. Then he explained the strategic advantage of being "meaningless," that other countries don't bother to get upset with Norway. At the end of the passage, he added to the construction of Norway as an insignificant nation by drawing on the sense of Norway as a "peaceful nation" (referring perhaps to the Nobel Peace Prize as well) and its key involvement in the Oslo Accords. In construing Norway as small, meaningless, and peaceful, he was constructing in our conversation a sense of Norwegian innocence that is very prevalent among Norwegians.

The phrase "Norway is a little land" is also drawn upon to emphasize that Norwegians can't fix everything. It gives the sense that they are doing "their part." On some level, this argument is certainly convincing. If every Western nation were to give one percent of its gross domestic product to humanitarian aid, as Norway does, things would probably be different. In this usage, the phrase turns blame back onto those who are "worse," especially the United States, as described earlier. It serves to imply that "the problem isn't really us. We, in fact, are innocent." It emphasizes the insignificance of widespread individual and collective behaviors among Norwegians that create climate gases. In this sense, Norway has also emphasized its cleanliness. Norway may be little, but the nation's ecological footprint, at 6.2 hectares per capita, is significant and one of the highest in Europe.[3] And, as noted earlier, the difference between its per capita greenhouse gas emissions and those of the United States is actually very small, even though Norwegians constantly portray the United States as the overfed hog emitting extremely large amounts of gas.

Table 5.1
Tools of Order and Tools of Innocence

Tools of Order
Maintaining conversational control: focus on facts, joking
Finding a *haldepunkt*: looking to the past, tradition, sense of place, connection to nature
Mythic Norway

Tools of Innocence
Mythic Norway
"Norway is a little land"
"We have suffered"
"Amerika" as a tension point

Not every usage of the phrase "Norway is a little land" is meant to exempt Norwegians from responsibility. The sense of being small is a part of how Norwegians understand themselves. It implies as well a sense of vulnerability and powerlessness. For example, Kåre Hagen and Jon Hippe write, "As a small country, with one of the most export-dependent economies among industrialized countries, the conditions for economic growth [in Norway] are largely determined beyond the reach of national polices" (1993, 101). And Lars Mjøset comments, "Norway is a country of about four million inhabitants. It is sometimes argued that small economies form a separate group since such countries are particularly strongly exposed to external challenges because of their economic openness" (Mjøset 1993, 108; see also Katzenstein 1985). Nevertheless, the general presence of this type of framing naturalizes the sense of Norwegian insignificance.

Claims to Virtue

According to Lifton, claims to virtue work to justify numerous actions that would otherwise be unacceptable. As noted earlier, two such claims to virtue were in use in Bygdaby and the nation as a whole with respect to climate change beginning in the early 1990s. Although the Norwegian government spoke urgently of the need to reduce emissions of climate gases, it was at the time involved in two projects that would do exactly the opposite: the building of two new natural gas facilities and the expansion of the petroleum sector by increasing oil development. The government justified both actions by switching the focus from national emissions-reduction targets and measures (as specified under the Kyoto Protocol) to an emphasis on climate change as an *international* problem and by attempting to meet Norwegian climate commitments through the *trading* of climate gas emissions rather than through the reduction of actual output. These arguments can be understood as national-level "impression management." They are examples of what Cohen (2001) calls interpretive denial.

Norway had by then taken early leadership on the climate front. In 1989, it was the first nation in the world to set a target for stabilization of carbon emissions. Not long afterward, however, it dropped this goal (Lafferty and Meadowcroft 2000). Eivind Hovden and Gard Lindseth describe how, instead, "the focus on national action to reduce GHG [greenhouse gas] emissions [was] replaced with an equally committed focus on the so-called Kyoto mechanisms and, more generally, the supposed positive international climate effects of the Norwegian petroleum industry. There is no Norwegian national target for reducing GHG emissions" (2004, 63–64). They further describe how

> the policy change that came about in the mid-1990s was made possible by a gradual discursive shift from the NA [national action] discourse to the TG [thinking globally] discourse. Only through the latter could Norway maintain both an expansive petroleum industry *and* international credibility in environmental matters. . . . For the NA discourse the petroleum operations represent a significant problem for Norwegian climate policy, whereas for the TG discourse the petroleum operations *are* a form of climate policy. Whether through the direct export of oil and gas, the direct export of gas-based electricity or . . . domestic use of gas-based electricity, the arguments of the TG discourse essentially revolved around the same line of reasoning: *since Norwegian petroleum products are internationally relatively clean, Norwegian oil and gas production is good climate policy internationally.* (74, emphasis in the original)

So the government changed the discourse it used: "By framing climate change as a global issue, local actors were able to portray the natural gas project as environmentally friendly" (Lindseth 2006, 739).

The Labor Party in Norway promoted the construction of the building of the gas plants on the grounds that because natural gas produces less carbon dioxide than sources such as coal, Norway could sell this excess energy to other nations and actually be helping overall global emissions. Thus, although the government acknowledged that Norway's emissions of climate gases must decrease, it used a claim to virtue to argue that by building two new natural-gas plants—thereby in fact *increasing* Norway's contribution to climate gases—it was actually helping to solve the problem of global warming. This strategy has been met with criticism: "While it is claimed that these [additional gas emissions] would be off-set by reductions elsewhere, this does not change the fact that emissions from Norwegian gas-based power would increase the CO_2 emission reductions that Norway would have to complete in order to fulfill its international obligations" (Hovden and Lindseth 2002, 158).

A second example follows a similar pattern. The government has argued that "since Norwegian petroleum products are not the dirtiest in the international market, Norwegian oil and gas production is good climate policy internationally" (Hovden and Lindseth 2002, 153). Norwegian researchers Hovden and Lindseth describe how

Miljkosok, an environmental cooperative forum consisting of the petroleum industry, the government and various interest groups and organizations produced a report in 1996 that in effect, concluded that Norwegian oil production was environmentally benign. The arguments were (a) that a cut in Norwegian production would increase the price of oil on the world market, which would make coal more competitive, and, most importantly, (b) that as Norwegian petroleum production has fewer emissions per unit oil produced, it was environmentally preferable to the oil produced by other countries. The unavoidable conclusion was that Norway should increase its Continental Shelf activity, as this would, in sum, be beneficial with respect to the global emissions of CO_2 and NO_x. (2002, 152)

Selective attention is an important part of Eviatar Zerubavel's (1997, 2006) analysis of the sociology of denial. And Rosenberg describes the importance of selective attention in the construction of reality. Part of the process of "imagining community" (Anderson 1991) is agreeing upon what people will collectively pay attention to (i.e., official national identity) and what they will collectively ignore (i.e., what happens, for example, on the other side of the Russian border, with environmental problems, and so on). When certain images are made "typical," they become overcommunicated, the focus of a disproportionate degree of attention. But if distraction were the only issue, it might not matter what alternate images or information captures public attention. Rather, images of national identity—in Norway's case, an identity based on equality, humanitarianism, proximity to nature, rural life, and simplicity—exist in a particular tension with environmental degradation and a high standard of living. In the face of contrary evidence, knowing and repeating facts about Norway (e.g., "Norway is number one in terms of per capita humanitarian aid"), making comparisons with worse conditions in other countries (the United States is a common target), and participating in symbolic rituals such as protest marches against racism are all ways to reinstate the threatened moral order. In situations of doubt and tension, elements of national identity can be put forward to negotiate the contradictions that Norwegians experience between Mythic Norway and modern realities, between their professed beliefs (practicing equality and humanitarianism) and their actions (spending money on luxury goods while neighbors or people in other countries struggle). The fact that one gave money to a project in Guatemala last week or sponsors a child in Romania can be offered up as justification if one feels guilt for purchasing a vacation package to Greece.

In table 5.2, I lay out the relationships between the "problematic" emotions of fear, helplessness, guilt, and "fear of being a bad person" or threats to identity that *Bygdabyingar* described in connection with the topic of global warming. Alongside these I list the extant emotion norms of being optimistic, maintaining control and national pride. The third column lists corresponding emotion management strategies I observed. I show how the emotions of fear and helplessness contradict the emotion norms of being optimistic and maintaining control. The negative emotions are particularly managed through the use of selective attention: controlling one's exposure to information, not thinking too far into the future, and focusing on something that can be done. Although the range of emotion management techniques indicated in the table's third column appear to be used across the community, I found that these strategies are used with more frequency by educators, men, and public figures.

Table 5.2
Relationship Between Emotions, Emotion Norms and Emotion Management Strategies with Respect to Global Warming in Bygdaby

Emotion	Emotion Norm	Emotion Management Strategy
Fear, helplessness	Be optimistic, maintain control	*Selective attention* Controlling exposure to information Not thinking too far ahead Focusing on something you can do
Guilt, identity	Be proud of Norway (managing spoiled identity)	*Perspectival selectivity* Not as bad as the Americans Norway is a small country anyway

Guilt and the fear of being a bad person contrast not only with specific local emotion norms surrounding patriotism, but also with the general social psychological need to view oneself in a positive light (i.e., manage identity). Guilt and negative self-identity are managed through the use of perspectival selectivity: by emphasizing Norway's small population size and the notion that no matter what Norwegians do, they are not as bad as the "Amerikans."

6
Climate Change as Background Noise in the United States

No longer an apocalypse ahead, critical environmental problems and constraints help construct society's sense of daily normality. Far from going away, environmental crisis has become a regular part of the uncertainty in which people nowadays dwell.
—Frederick Buell, *From Apocalypse to Way of Life*

While 61% of Americans say the effects of global warming have already begun, just a little more than a third say they worry about it a great deal, a percentage that is roughly the same as the one Gallup measured 19 years ago.
—Frank Newport, "Little Increase in Americans' Global Warming Worries"

I think it's hard to make people change, especially Americans. We have a very strong mythology about the United States. Since we are different than everybody else, smarter, have more initiative, that we're basically exceptional.
—Cindy, concerned American

I have examined the lived reality of public silence in one Norwegian town at the beginning of this century, looking at how climate change was present in people's minds but in a fleeting and unfocused manner, observing differences between front-stage invisibility of the problem and backstage concern about it, and analyzing in some detail how people used culturally available tools to manage emotions and re-create a sense of normalcy and innocence in the face of their unease. To what extent are these reactions replicated around the world? Public silence in the face of climate change is not unique to this one town in Norway. Indeed, there is evidence that people around the world experience deep fears regarding climate change, struggle to make sense of guilt, and in many cases normalize their inaction through a variety of cultural tools and narratives.

Social commentators across Europe describe publics who are "sleepwalking," "apathetic," or "in denial." Michael McCarthy, environmental editor for *The Independent*, writes that "global warming is being greeted with a yawn by half the population of Britain" (2007). Project Omelas is dedicated to "overcoming apathy in Australia." Indeed, the language of denial and apathy regarding climate change are now widely used in cyberspace on blogs, listservs, and all forms of online commentary. *The Independent*'s staff writer Geoffrey Lean writes that "future historians, looking back from a much hotter and less hospitable world, are likely to play special attention to the first few weeks of 2005. As they puzzle over how a whole generation could have sleepwalked into disaster—destroying the climate that has allowed human civilization to flourish over the past 11,000 years—they may well identify the past three weeks as the time when the alarms last sounded" (2005). Renee Lertzman from Cardiff University writes in *The Ecologist* that "public apathy is fast becoming one of the hottest topics in environmental circles. It would appear that people do not seem to care or be moved to action in the face of urgent ecological threats. Running a close second to apathy is the topic of denial; the stunning way in which people can literally deny or pretend things are not as they are, creating enormous barriers and psychological blocks for making necessary change" (2008, 16; see also Lertzman 2009).

Among the places in the world where public inaction matters most is the United States. The combination of high U.S. carbon emissions and American political and economic power underscores just how much is at stake for U.S. involvement. And cultural hegemony places Americans in the spotlight at the same time as it creates a powerful political opportunity for American citizens. In a talk to the American Sociological Association in August 2005, Indian writer Arundhati Roy referred to all these conditions when she told her audience that, "as Americans, you are living on the palace grounds" and that American privilege gives Americans a special responsibility. Life on these palace grounds may not always be "easy," but ultimately the complicity or engagement of the American public on climate change has powerful consequences for present and long-term human well-being and the ecology of the planet.

Having been the subject of significant academic analysis, criticism from the environmental community, and outrage from other nations around the world, American failure to engage with climate change is well established. In July 2009, the University of Maryland's Program on International Policy Attitudes released a comparative study of 19 countries that showed Americans were near the very bottom of the pack in their sense of how much emphasis their government should put on climate change. Even Forbes.com featured a story titled "American Apathy and Global Warming." Yet part of why a case study set in Norway is so valuable is that as we turn our attention to the United States, things rapidly get a whole lot more complicated. Public silence in the United States has been both similar to and different from what I have described in Norway. Whereas in Norway we speak of a public silence in the face of known (and largely accepted) information on climate science, in the United States newspaper readership and general literacy regarding climate change are demonstrably lower than they are in most industrialized nations.[1] Vast internal differences in economic resources, political orientation, and geography bring further complexity to any discussion of American silence. And American rates of participation in political parties and elections and of civic engagement in general are notoriously lower than their counterparts around the world (Niemi and Weisberg 2001; Putnam 2000). Why should we expect Americans to respond to climate change, when they fail to engage in anything else?

Furthermore, when it comes to American public opinion on climate change, everyone wants to talk about skepticism regarding global warming. The percentage of Americans who question climate science is among the highest in the world (GlobeScan 1999). And the fact that the percentage has increased sharply just in the past few years has been the focus of much scholarship and media attention. Why is it that fewer Americans trust climate scientists? What does such skepticism say about the place of science in American public life? And why has the media portrayed climate change as controversial even when there is widespread consensus on it within the scientific community? Such have been the important quandaries upon which American social scientists have focused their attention (see especially Jacques 2009, but also McCright and Dunlap 2000, 2003; Boykoff and Boykoff 2004; Boykoff 2008a, 2008b; Jacques, Dunlap, and Freeman 2008). In this chapter, however, I explore how even or perhaps especially in the United States these more overt forms of skepticism (again, what Stanley Cohen [2001] calls "literal denial") have overshadowed another larger and more insidious phenomenon—the way a majority of the public is concerned but has normalized their knowledge rather than acting on it.

The events of Hurricane Katrina in August 2005 certainly illustrated that the United States will not be immune to the catastrophic impacts of large-scale storms caused by global warming. Even given that we cannot link this specific weather event to climate change, the events surrounding the initial impact and response effort to Hurricane Katrina provided the opportunity to provoke public discussion about what future climate scenarios might mean for our country. Sixteen months after Katrina, climate change finally made it onto the Pew Research Center's (2009) measures of the top ten public priorities for national action. This event, together with the widespread viewing of Al Gore's documentary *An Inconvenient Truth* in 2006, marked the turn to an era in which climate change at last became a publicly acknowledged problem in the United States. Physicist and author Spenser Weart, who has written on the history of climate change in the United States, notes that "attention to climate change in the American press climbed to the highest level ever during the Fall of 2005." He writes that

In November 2005 alone, PBS public television stations, the Turner Broadcasting System, and even the right-wing Fox News Channel all ran specials stating plainly that global temperatures would rise, and a much larger audience saw movie idol Leonardo DiCaprio explain the problem on the *Oprah Winfrey Show*. The Weather Channel added reports on climate change as a "niche" market. In the spring of 2006, people could see a thorough analysis of the danger in two widely read books by top science journalists, a week-long series of reports on ABC television and radio, and a special issue of *Time* magazine ("Be worried," the cover advised. "Be very worried.") (forthcoming)

By 2006, climate change had finally reached the status of a "real" problem in the United States, but it continues to remain invisible in daily lives and in government planning, just as it has been for the people of Bygdaby and for the Norwegian government. The Norwegians we have been following recognized the seriousness of climate change as a problem but failed to integrate such understanding in their personal lives or political decision making. And now that a majority of Americans also recognize the seriousness of climate change, they also appear to face the difficulty of coming to terms with it head on. Yale scholar Anthony Leiserowitz, who has conducted extensive work on public opinion on climate change, notes that "large majorities of Americans believe that global warming is real and consider it a serious problem, yet global warming remains a low priority relative to other national and environmental issues and lacks a sense of urgency" (2007, 44). Similarly, in writing up the summary of Gallup's 2008 environment report on the Gallup Web site, staff writer Frank Newport notes, "Despite the enormous attention paid to global warming over the past several years, the average American is in some ways no more worried about it than in years past" (Newport 2008).

Denial in the United States matters both because of the aforementioned U.S. cultural, political, and economic hegemony and because it provides a context to examine the relationship between it and skepticism regarding global warming. Particularly in the United States, literal and implicatory denial go hand in hand. Although on the face of it these two trends, skepticism and denial, appear quite different (after all, in any public debate they would be on opposite sides, perhaps even exchanging heated words), they are related. The fact that nobody wants information about climate change to be true is a critical piece of the puzzle that also happens to fit perfectly with the agenda of those who generate skepticism. There is an important congruence between these troubling emotions and the psychological defenses they engender, on the one hand, and the social structural interests in minimizing public responses to climate science, on the other. Although true in Norway, this relationship is even clearer in the United States, where a massive, well-organized, and well-funded campaign of skepticism regarding climate change has been carried out (McCright and Dunlap 2000, 2003; Jacques, Dunlap, and Freeman 2008; Jacques 2009; see also Michaels and Monforton 2005 for earlier history of antienvironmental industry tactics). The facts that people find climate change difficult to think about and, indeed, that nobody wants information about climate change to be true happen to fit perfectly with the agenda of those who have put forward the messages of skepticism in the media. These facts create exactly the kind of slippery condition that makes implicatory denial feel so natural. A large group of the public who would otherwise have felt accountable to act can latch onto the possibility that climate science may not in fact be true as a justification for their failure to act. And even for those who clearly believe climate scientists' conclusions and are worried, the presence of discourses of skepticism in the public sphere reduces the apparent stakes of the issue by making the matter appear to be about opposing political viewpoints rather than about human survival.

Climate change is significant in its size, scale, and scope as well as in the degree of threat it poses to ecological systems and human society. But from the standpoint of human social behavior and modern political theory, climate change is significant in two additional important ways. The two citizen responses of skepticism and denial—that is, some know about climate change but manage to consider as no more than background noise the possibility that life as we know it will end, and some do not believe that science should be the basis for guiding public policy—are unique in modern history. No previous environmental problem has generated either response with such force. Furthermore, in very different ways each type of response flies in the face of basic assumptions regarding human behavior that go back to the Enlightenment and the origins of modern society. Each of these responses poses unique threats to democracy and unique challenges for social theorists and public commentators (see Jacques 2006 and Jacques 2009 for an excellent discussion of the history and political significance of skepticism regarding global warming).

A brief look at the larger terrain of American perceptions of climate change offers some interesting insights. Although a new public-opinion survey reports on the same set of questions regarding climate change every month, there are frustratingly few examples of data in the form of in-depth interviews or focus groups that might provide insight into what is going on behind the numbers. This chapter draws on a recent set of interviews on climate change in a rural community, national-level survey data, and observations from other scholars to establish parallels and show differences between the Norwegian and U.S. cases. Comparisons with the United States also include the voices of some of my students as they address both the emotions they feel regarding climate change and why and how they manage those emotions. These voices, although less systematically collected, are offered for the light they may shed on the broader resonance of the experiences of people in Bygdaby as well as for what they show about differences across national contexts. This chapter is organized to follow parallel examples of the themes from chapters 3, 4, and 5, albeit with a much smaller sample of voices. That is, I touch on the disturbing emotions that people articulate, illustrate selective attention and conversational control as emotion-management techniques, and, finally, offer a few examples of social narratives that comprise the uniquely American tool kit for dealing with the issue of climate change.

Climate Change in the United States: Important Scientific History but off the Public Radar

I feel like in my circle of friends I'm always talking about how I'm nervous about the future 'cause I don't know what is going to happen. But I have a lot of friends who have this mindset that they are not going to be affected.

—Teresa, American college student

I know for several years we've had lower snow pack and river levels than previously. . . . I imagine that we're going to see some issues maybe with water availability. I think there is probably a higher likelihood of a forest fire up in the

city's watershed, which has a huge fuel load, anyway . . . but I think it's uncomfortable to think of things changing and about changing our own way of living in order to prevent something that's not immediately perceived as knocking on our door.
—Cindy, rural American city council member

Public indifference to climate change and the apparent perception that climate change is not worth worrying about stand in dramatic contrast to the fact that the United States has been on the forefront of climate science at least since Dr. Stephen Schneider and others at the National Center on Atmospheric Research produced world-renowned work on the issue back in the 1980s. Climate change was first a front-page story in the *New York Times* on October 18, 1983, when the U.S. Environmental Protection Agency released a report that the earth would begin to heat up in the 1990s. Five years later, in 1988, NASA scientist James Hansen testified before Congress on the dangers of climate change. That same year was marked by the establishment of the Intergovernmental Panel on Climate Change (IPCC). Two years later the first IPCC report was released with the announcement that the globe had been warming and that future warming was likely.

Each of these events was an opportunity for people to reflect on the significance of climate change. These events were widely reported in the U.S. newspapers. I even took a semester-long class on climate change as an undergraduate student in 1989. In 1992, the Earth Summit in Rio de Janeiro, Brazil, resulted in the creation of the United Nations Framework Convention on Climate Change, which in turn led to the Kyoto Protocol in 1997. Meanwhile, the second IPCC assessment report was released in 1995 and emphasized the human origins of climate change and the seriousness of future warming. Scientific research on dramatic melting of glaciers and Antarctic ice shelves was also widely reported in the news during the mid-1990s. By 2001, with the release of the third IPCC assessment report and the report from the panel of the National Academy of Sciences Committee on Abrupt Climate Change, debate among all but a few scientists had ended, marking a situation in which there was as much or more consensus on climate change as on any contemporary scientific discovery.

Yet despite this degree of scientific consensus and despite the location of the United States as a prominent source of climate science, this country surely represents the most pronounced example of public silence on the topic worldwide. Although the United States has exhibited scientific leadership on climate change, the issue had nearly complete invisibility in daily life, government planning, and mitigation activities on the local or national agenda until well after the year 2000. A Government Accountability Office report from 2007 charges that federal agencies have been slow in incorporating climate science in their strategic management planning: "In particular, BLM, FS, FWS, NOAA, and NPS have not made climate change a priority, and the agencies' strategic plans do not specifically address climate change," and although "a broad order developed in January 2001 directed BLM, FWS, and NPS to consider and analyze potential climate change effects in their management plans and activities, the agencies have not yet provided specific direction to managers on how they are to implement the order" (U.S. Government Accountability Office 2007, 2).

For evidence of this disjuncture among the public at large, one need look no further than the titles of stories released by Gallup on its Web site reporting the results of its annual Environment Poll:[2] in April 2001, "Americans Consider Global Warming Real, but Not Alarming"; but three years later, in March 2004, "Global Warming on Public's Back Burner." Even after Katrina hit, the March 2006 Gallup headline read "Americans Still Not Highly Concerned about Global Warming." These stories were followed in subsequent years by stories with such titles as "To Americans, the Risks of Global Warming Are Not Imminent" (March 2007) and "Little Increase in Americans' Global Warming Worries" (April 2008). How is it possible that predictions of major threats to social infrastructure such as sea-level rise or dangerous and costly impacts from increased wildfires and flooding failed to mobilize public response? Those who evaluate security risks on a professional basis may have taken note, but the category of climate change did not even make it onto the Pew Research Center's annual January list of national domestic priorities for the government and Congress until 2007—some 24 years after the first front-page story on climate change in the *New York Times*! And not until 2008, a full 20 years after Dr. Hansen's congressional testimony, did the economic, ecological, security, and justice implications of increased drought, wildfire, and storm activity from climate change merit attention as a presidential campaign issue.

Although Norway's location in the far north may, on the face of it, enhance the Norwegian sense of the immediacy of climate change in daily life, communities in the United States have also begun to face climate impacts on economic, social, and cultural fronts. Parts of the United

States are, like Norway, in the far north, where climate impacts are intensified. The Arctic is warming at nearly twice the rate of the rest of the planet. As permafrost melts, Alaskans have seen homes and highways shifting and lakes disappearing (Wendler and Shulski, 2009). Entire island communities have been relocated owing to decreased sea ice. Arctic indigenous people face particular impacts (Agyeman, Cole, Haluza DeLay, et al. 2009). As Inuit Nobel Prize nominee Shelia Watt-Cloutier explained in testimony before the Inter-American Commission on Human Rights,

> For Inuit, sea ice allows for safe travel on the perilous Arctic waters and provides a stable platform from which to hunt its bounty. The ice is not only our "roads" but also our "supermarket." Deteriorating ice conditions imperil Inuit in many ways. Ice pans used for hunting at the floe edge are more likely to detach from the land fast ice and take hunters away. As the ice is melting from below, hunters can no longer be certain of its thickness and how safe it is to travel upon. Many hunters have been killed or seriously injured after falling through ice that was traditionally known to be safe. . . . These impacts are destroying our rights to life, health, property and means of subsistence. States that do not recognize these impacts and take action violate our human rights. (2007)

University of Alaska economist Peter Larson and his colleagues recently estimated that damages to Alaskan infrastructure from climate change may cost the state $10 billion over the next few decades (Larsen, Goldsmith, Smith, et al. 2008). Economic costs also extend to communities across the United States from mountain zones that face the loss of ski industries (Hamilton et al. 2003; Scott, McBoyle, and Mills 2003; Scott and McBoyle. 2007) to coastal regions that confront increased damage from sea-level rise and storm surge. For example, a Florida State University report released in 2008 found that climate change will cause significant impacts on Florida's coastlines and economy due to increased sea-level rise (Harrington and Walton 2007). Economic costs from damage associated with extreme storm events will be considerable—$6.7 billion by 2080 in Dade County alone. Recent changes in precipitation patterns and snowpack across the Western United States have led to significant costs in the form of agricultural losses, increased wildfire, and impacts to local water and power supplies (Frederick and Major 1997; Barnett, Adam, and Lettenmaier 2005).

Industries who evaluate risks on a professional basis have made a point of heeding scientific studies and have led the way in calling for precautionary approaches. The May 2007 testimony of Franklin W. Nutter of the Reinsurance Association of America on the economic impacts of global warming before the U.S. House Select Committee on Energy Independence and Global Warming is quite revealing. As Mr. Nutter explained, "No financial services business is more dependent on the vagaries of climate and weather than property and casualty insurers." Mr. Nutter testified that insurance damages have significantly increased in recent years, creating a situation about which the industry is very worried. He also noted that although U.S. insurers represent the U.S. industry that has taken the most leadership in integrating scientific climate assessments into their daily practice, they still lag significantly behind their European counterparts:

Although a number of European insurers and reinsurers have shown great interest in understanding the causes of climate change, including the impact of global warming, U.S. insurers have been more focused on the effect of extreme weather events. Thus, the U.S. industry has been more attentive to approaches to mitigate the consequences of natural catastrophes and other extreme events; while some European insurers have called upon their governments to reduce the human factors they believe contribute to global warming. (Nutter 2007)

Another institution in charge of evaluating security is obviously the military. President George W. Bush publicly questioned climate science and declared it not in the American economic interest to participate in Kyoto, but within the Pentagon there was clear concern. In 2003, the Pentagon produced a report to President Bush describing climate change as a "threat to global stability [that] vastly eclipses that of terrorism" and concluding that climate change "should be elevated beyond a scientific debate to a U.S. national security concern." The report, *An Abrupt Climate Change Scenario and Its Implications for United States National Security*, forecasts widespread rioting, the possibility that major European cities will disappear with sea-level rise, that Britain may be plunged into a "Siberian" climate by 2020 and the risk of increased nuclear threat as countries attempt to defend supplies of food, water and energy. The authors of the Pentagon study conclude that "it is quite plausible that within a decade the evidence of an imminent abrupt climate shift may become clear and reliable. . . . In that event the United States will need to take urgent action to prevent and mitigate some of the most significant impacts" (Schwartz and Randall 2003, 22).

Yet somehow these concerns have failed to gain the degree of political or social traction such serious threats would seem to warrant. Thus, in the United States, just as in Bygdaby, we can also find evidence of a "double reality," a widespread awareness of climate change coupled with its invisibility in conversations, political organizing, and planning processes. Most Americans are finally convinced that climate change is happening. In fact, in 2008 more than 40 percent of us even believe its consequences will pose a serious threat to us in our lifetimes. Yet somehow this same group continues to see climate change as no more than background noise. Consider these statements in the write-up from Gallup's 2008 Environment Poll:

> Despite the enormous attention paid to global warming over the past several years, the average American is in some ways no more worried about it than in years past. Americans do appear to have become more likely to believe global warming's effects are already taking place and that it could represent a threat to their way of life during their lifetimes. But the American public is more worried about a series of other environmental concerns than about global warming, and there has been no consistent upward trend on worry about global warming going back for two decades. Additionally, only a little more than a third of Americans say that immediate, drastic action is needed in order to maintain life as we know it on the planet. (Weart 2009)

Yale climate scholar Anthony Leiserowitz links the lack of urgency regarding climate change with Americans' "[perception that] climate change [is] a moderate risk that will predominantly impact geographically and temporally distant people and places" (2005, 1433). Such statements appear to be obvious explanations on face value, but if so then how do Americans perceive what is happening in Alaska, where climate change impacts are particularly evident, and why do they remain mainly skeptical? If we take Leiserowitz observation of public experience as an explanation for public behavior, we miss the opportunity to ask, "How are near and far constructed in particular thought communities?" And it is even more important to ask, "Why are near and far constructed as they are?" In short, we skip over the politics of emotion and knowledge; we omit the many important contributions that may come from the field of psychology, and we pass over the possibility that power might be operating or reproduced in constructions of relevance and irrelevance. We miss most of what is unique, complex, or new about climate change as a social problem. To explore these dimensions further, I turn again to qualitative data for views behind the scenes of such behavior in the United States.

Fear, Helplessness, and Guilt in the United States

In Bygdaby, we saw how inaction was linked behind the scenes to troubling emotions, including guilt, hopelessness, and concern about the future that community members faced. To what extent are such emotions relevant in the United States? Americans, too, experience fears about the severity of climate change's long-term impacts and about whether their way of life must change, and they feel both guilt regarding their actions and a general loss as to how to respond. Although voices from Norway find resonance in the United States, unique features of the American political and cultural landscape add additional challenges for those who would come to terms with climate change.

Fear

There's a lot of uncertainty that comes with changing weather. As far as growers are concerned, you know, how do I anticipate the growing conditions when they are going to change? Will trees I plant be able to survive by the time they mature, you know, the conditions might be different. That's a worry. And this valley is largely agricultural. People will have to adapt, and if there are significant shifts, it will be more difficult to adapt.
—Jeff, American farmer, midthirties

Near my hometown, several wheat farmers gathered in a downtown café one morning in early March to do an interview about climate change. These men were longtime residents of the region, in their fifties and sixties, politically engaged, and active in their larger community. Their views on climate change were varied, reflecting a mixture of curiosity, concern for the future, and perhaps some disguised skepticism about global warming in polite deference to the obviously middle-class, urban, college-educated interviewer who sat across the table from them. Regardless of whether they thought that climate change was human caused or that Al Gore's film was a bit over the top, they remarked on the intimate importance of climate to their livelihoods and their clear sense that things were changing. One farmer spoke about water: "We've seen things we've never seen before, and we've been here all our lives. And to see dust in January, it almost reminded you of an August dust storm. I've never seen that in January. . . . And I've talked to some of the men who go out and probe the soil; they say it's dry a couple feet down. It's kinda scary. You'd think that with all the snow and everything we've had, but we really don't have a lot of submoisture yet. That's kind of scary."

A few miles away and on the other end of the age and political spectrum, during public lectures, films, and class discussions related to climate change my college students described emotions of anger, guilt, helplessness, and fear for the future. When I asked students in my environmental social movements class whether they had ever experienced fear about the future in relation to climate change, the reply was a unanimous "yes." One young politically active woman commented: "When I heard Bill McKibben speaking, and they talk about the tipping point, fear comes in with helplessness, 'Oh my god, there is nothing. . . .' I found out, you know, we're 50 years away from the worst; we haven't even felt the effect of what we are doing now. It makes me think, 'We can't do anything, it's already beyond help,' and then there's the fear." Students spoke about these fears in vivid detail, interrupting one another as they described how they had very little conception of what could be in store for them, not just in the next 15 years, as asked, but even one year out:

Jocelyn It's kind of like a science fiction movie. We're, like, *who knows*? Hurricane Katrina and things like that could be all over everywhere; we have no idea.
Robin Like what was that movie? *The Day after Tomorrow*. That is what I'm afraid of.

Students speak rapidly now, interrupting one another with nervous laughter:

Joey Yeah, I'm terrified of what's going to happen *next* year. (Laughing with embarrassment.)
Mary It's like everything comes together in this crazy moment, and all these storms happen all over the world and tornados and earthquakes.
Teresa And the Northern Hemisphere is blanketed with snow!
Lauren Mexico closes its borders to us!
Jocelyn No, it's terrifying. (Speaker is serious, background laughter fades out.)

These students voiced fear in the context of whether the problem can even be solved or not. One young woman specifically noted that fear came up for her "when thinking about if current efforts are going to be sufficient." Their general sense of fear for the future was also linked with the second emotion highlighted by the people in Bygdaby: helplessness.

Helplessness

A sense of unresolved, perhaps irresolvable, environmental crisis has become part of people's normality today. Faith in effective action has diminished at the same time that the concern about the gravity of the crisis has sharpened.
—Frederick Buell, *From Apocalypse to Way of Life*

A second emotion confronted by people in Bygdaby in relation to climate change is helplessness. People in that community described how large the problem seemed, not to mention the need for the cooperation and common work of people in so many different countries or the fact that entire economic structures will likely have to change significantly if we are, for example, to reduce emissions by 60 percent or more in the next fifty years, as is widely recommended. In the face of what is now commonly called a "superwicked problem" (a highly complex social problem for which, among other things, there is no clear solution, a need for extensive change, and the sense that time is running out), it is hardly surprising that people feel helpless. These expressions of helplessness are certainly echoed in the United States, and I encounter them every day in my teaching. One community member I interviewed, a young man in his early thirties, described helplessness in light of the enormity of climate change: "I do feel somewhat helpless because as much as I can talk about people being apathetic and not doing anything and taking responsibility for their actions, it takes a lot of energy just to survive in whatever context, you know, U.S. or elsewhere. It takes a lot of energy. So I'm tired. So I do feel helpless. That I don't have enough energy to help."

An American Geophysical Union focus group study similarly links declining concern to a sense of hopelessness and frustration. The authors emphasize the public's feelings of powerlessness and frustration connected to the issue of climate change (Immerwahr 1999). In reflecting on her lack of engagement, one of my students commented on how "solving global warming can seem like such a daunting task, and even I know that it can seem too overwhelming." Mary, a female student, told our social movements class, "It's often hard to think about the mass amounts of problems and then thinking about how little I actually do to resolve or minimize these problems." Another student, Lauren, added, "Even after hearing activists talk about their work, I still feel removed from it and helpless."

Helplessness is a powerful emotion. Bear in mind the work of psychologists on the desire people have to feel in control of their worlds (a sense of self-efficacy) as well as the application of notions of cognitive dissonance to American perceptions of climate change. Recall in particular how Jon Krosnick and colleagues (2006) found that many people judge as serious only those problems for which they think action can be taken. As was vocalized regarding the emotion of fear, a key dimension of the helplessness was linked to a feeling that existing political structures are inadequate to handle a problem as large as climate change or are impenetrable to the interests of everyday people or both. Brett, a male student, described "a feeling of impotence," and Ted noted how getting information without what he perceived to be adequate options for action has led to paralysis:

> I think that we ran into this at the [E-Law] conference [University of Oregon, Eugene, 2009]. We'd hear all this information and get all riled up, and then they'd be, like, "contact your legislator," and I'd be, like, "Aw, really? Can't you give us a little more?" That's, like, where I feel the most helpless; it's like I know all this stuff, I have all this information, [but] what the hell do I do with it? I don't know where to turn, I don't know who to talk to. Yeah, I can write my congressman a letter, but in all honesty . . . I am not sure that one person can make such a difference.

Note that here, for Ted, helplessness is linked with both his questioning that "one person can make such a difference" and a lack of confidence in the political system. His concerns highlight a unique dimension of the American experience of climate change. Unlike in Norway, where political affiliations and engagement are widespread, in the United States the general sense of helplessness that comes from facing such a large problem as climate change is exacerbated by the pervasive culture of individualism (Maniates 2002; Bellah, Madsen, Sullivan, et al. 2008; Szaz 2007), extensive political alienation (see, e.g., the classic work by Putnam 2000), and a lack of fluency in the workings of the political system that contribute to people's sense of political empowerment and their actual ability to effect change. Gallup surveys of Americans' trust in their political leaders, for example, indicated that only 2 percent "just about always think they can trust the government in Washington to do what is right" (Gallup News Service 2010). Rates of voter participation in Norway are among the highest in the world; the same cannot be said of the United States. Moreover, climate change challenges our existing political and economic structures like no other prior problem. As Anthony Giddens notes in his recent book on the topic, a key problem is that "we have no politics of climate change" (2009, 4), by which he means that we have no real analysis of how to reduce our emissions to necessary levels. The authors of the American Geophysical Union study likewise conclude that public apathy hinges on whether anything can be done: "As we have said earlier, informing the public of the problems can increase frustration and apathy rather than build support. Our research suggests that what the public is most skeptical about is not the existence of problems but our ability to solve them. What will make the public invest energy in these issues is not the conviction that the problems are real, but that we can do something about them" (Immerwahr 1999).

In the United States, the sense of helplessness in the face of climate change is not only exacerbated by political alienation, but compounded by the culture of American individualism. American individualism is hardly new; after all, Alexis de Tocqueville emphasized this aspect of American character during his famed visit to the United States back in the 1830s. In their landmark study on the topic, Robert Bellah and colleagues have more recently emphasized the centrality of individualism in the American psyche. They write that "individualism is the first language in which Americans tend to think about themselves." Indeed, these authors find individualism evident "at every level of American life and in every significant group" (2008, xiv). They are critical of American individualism, however. They warn that American individualism is so pronounced in the United States as to create a "crisis of civic membership" in which "more and more of us doubt whether we can trust our institutions, our elected officials, our neighbors, or even our ability to live up to our own expectations for our lives" (xiii). Many important critiques of individualism highlight the loss of political power that comes from neglecting to unite with others and the particular translation of individual responsibility into consumerism such that people come to see themselves as consumers rather than as citizens (e.g., Maniates 2002; Szaz 2007). Regarding climate change, Jennifer Kent notes that "whilst there may be a compelling case for individual responsibility, not the least being the seriousness of the problem at hand, the very significant reductions in greenhouse gases needed, and the willingness of people to play their part, a reliance on personal contributions to greenhouse gas reductions may hinder the development of effective global policy and action whilst diverting public attention from engaging fully in civil society" (2009, 144). Moreover, as Kent highlights, individualism falls short in being able to combat climate change specifically because people sense that climate change is not a problem they can fix with individual action, especially when that action is limited to acts of so-called environmentally responsible consumption. Americans are so immersed in the ideology of individualism that they lack the imagination or knowledge of alternative political means of response. Because climate change requires so much more than individual action, discourses of individual responsibility, rather than enhancing agency, merely "[alert] individuals to their essential ineffectiveness in tackling complex global environmental issues"

(Kent 2009, 145). Thus, helplessness is the product of a situation where people perceive that neither individual actions nor existing political or economic structures are adequate. Kent notes that in the context of the recent emphasis on individual responsibility for climate change,

> feelings of powerlessness are further entrained as people realize their inability to effect global change through their individual agency, calling on their governments to act. . . . However people also understand global degradation as symptomatic of weak political action . . . so not only do individuals perceive an unacceptable level of action from governments on climate change mitigation [but] they are also cynical that governments are genuinely serious about climate change as it is understood to be against their economic interests. (2009, 143)

Bellah and his coauthors indicate another factor behind helplessness in the United States: a declining sense that one's fellow citizens care and can be counted on to respond. Among the U.S. public, the assessment that Americans are ignoring the issue because they are simply too greedy to care about climate change is not uncommon. Individuals see that smart people around them continue to carry on highly consumptive behaviors, even in the face of knowledge of their consequences, and assume that others are too self-interested to be motivated for change. My environmental studies students expressed this sentiment on a regular basis. There is also evidence of this perspective in the focus groups on climate change conducted in 1998 by the American Geophysical Union:

> Many respondents in our focus groups were convinced that the underlying cause of environmental problems (such as pollution and toxic waste) is a pervasive climate of rampant selfishness and greed, and since they see this moral deterioration to be irreversible, they feel that environmental problems are unsolvable. As a result, convincing people of the seriousness of the problems is at best only part of the solution, and may, in fact, be counterproductive. (Immerwahr 1999)

We are overwhelmed because we recognize the enormity of the problem but have no clear sense of what can be done and do not know whether other people also really care, whether the political system is up for the task, and whether their attempt to respond will generate even further problems! One of my students, Brett, explicitly addressed the sense that not knowing what to do is a barrier to action: "I think that one of the reasons I'm not as active as I'd like to be is that I'm waiting for this golden bullet solution, and I feel like all the things that are available to me now are not sufficient. And that's my reason for not participating."

Finally, helplessness, for my students, was linked not only to the enormity of the problem, but also to the power of industry and ethical questions about how to respond in ways that don't create further hardship in the global South.

Megan   For me, global warming is so much more than an environmental issue. I understand that we have to cut back on our emissions, and that's great, but what about industrializing societies? There's going to be huge emissions if they industrialize, but it's not necessarily right to keep them in a state of perpetual undevelopment to cut down on global warming 'cause we screwed up. So how do you reconcile those two things?

Julie   Yeah, I think that's where the helplessness comes in for me because I can do these things on my own every day, like walking, making those choices, but the big environmental problems are coming from industry, and I feel like I can't do anything about that.

Here both knowledge of American global economic privilege and the lack of a critique of industry (which is, again, a unique part of the American political landscape) contribute significantly to the sense these young people had about whether anything can be done.

Guilt

A third key emotion that emerged from the Norwegian ethnography is guilt. Recall how *Bygdabyingar* described awareness of their involvement and the way this knowledge made them feel guilty. In the Norwegian context, guilt is exacerbated by the combination of knowledge about high national emissions and oil wealth and the particular moral imperative defined by the Norwegian sensibility. Recall, too, that guilt, like a desire for self-efficacy, is one of the emotions that motivates cognitive dissonance. In the United States, the particular cultural, political, and economic contexts for emotions are of course different than in Norway, although in the United States knowledge of the impacts of our lifestyle on people around the globe is all the more confounding. The United States has until just recently been the largest contributor to climate change, accounting for approximately one-quarter of the total global greenhouse gas output, yet we rank third worst internationally in terms of our response, after China and Saudi Arabia. Per capita U.S. carbon emissions are also among the highest worldwide. These facts, especially when taken together with our dominant role in global politics and our sizeable economic resources, intensify the moral implications of our failure to act. Each of these structural conditions enhance guilt.

Young people in the United States also describe guilt in great detail. In fact, guilt is probably the emotion that my students expressed most vividly. As Teresa put it, "It's frequently uncomfortable because thinking about environmental problems is accompanied by knowledge of my contribution to the problem and frequently, too, [by] a lack of willingness to give up luxuries." Yet although some emphasized guilt in relation to not wanting to give up luxuries, many described guilt for participating in a system that they did not know how to escape. In this sense, guilt and helplessness are intertwined. According to Jocelyn, "I find that there is a kind of guilt because the way we live, culturally the way we've been raised, is so contrary to our ideals in a lot of ways. You can be against global warming to a certain extent, but we're still heating our homes and driving our cars and shipping our food long distances. It's really hard to live in a social context and be aligned with your beliefs on the environment." Barbara, another young woman, agreed:

It's hard to maintain any sort of—*righteousness* is not the word that I want—but to really feel that you can accurately represent your ideals and vocalize them; you kind of feel guilty when you do that. If you reflect on your actions throughout the whole day, you realize you drove to school, and then I got a to-go salad for lunch. I think a lot of people at our college believe that climate change is happening and learning about it in academic settings, but we're living these lifestyles where the heat in the dorms is blasting all the time, and we're using lights even when it's light outside, like right now.

Students remarked with emphasis on the overwhelming nature of the guilt that comes from being unable to escape participating in the social norms that have led to their nation's high emissions. Jocelyn described the feeling that no matter how hard she tried, she and everyone else will still go to "ecohell": "I feel like that guilt can be really overwhelming when all you hear about is what we're doing wrong. It's like we can maybe cut down on this or that, but you're still going to ecohell or whatever." At the same time that they don't know what to do or how to get out of the system, several students pointed to the source of their guilt as being the fact that they aren't doing more. Here again is Lauren: "And I feel guilt not just in environmental issues, like I'm worsening climate change or different things, but also I feel mostly guilty for not participating or being active. I shy away from political things; I might agree with a lot of initiatives but don't find myself leading them or acting. So most of my guilt comes from knowing about this and not doing much about it."

These young people expressed a profound sense of entrapment. Caught between their desire to do the right thing and to be informed (which they felt is a necessary first step toward responsibility and change), on the one hand, and the fear, guilt, and helplessness that has emerged in the absence of easy answers or perhaps of any solution at all, on the other.

Brett I was going to say I have similar feelings as Barbara. A lot of the times I do feel that I've *heard* everything, and you get to a point where you're like, wow, I need to stop learning about the details. You want to do something, like I'm sick of going to talks because it's another form of apathy. I'm just sitting here and listening to someone tell me that there's this huge problem, but nothing's changing, and I've heard this five years ago. Like I could have told you most of this when I was at the end of middle school, and nothing's changed. And so I think that I do have similar sentiments. And I don't go to as many talks, or I'll go, and partway through them I'm sitting there thinking, "Why am I still here? I know what he's going to say. And I've seen the little charts that he's going to put up."

Barbara (Laughing.) The hockey stick graph!

Jocelyn (Laughing too.) I've seen the little lines going up, and the graph getting really steep.

Brett I was going to say that I agree with you. I mean, a lot of time I go to talks out of a feeling of impotence. I mean, I can't do anything, so I might as well do this. And the science hasn't changed. It's gotten a little more precise, maybe there's going to be a few more effects that they will teach me about, but we know what we know, and we need to act on it.

Lauren That's something I was thinking about earlier, too; if you're trying to think of what you could do, even on a personal level, sometimes I'm unsure.

Barbara (Spontaneously taking the conversation in another direction.) I just want to ask something from the group because I wonder, like, how even having this discussion is making everyone else feel. Because I feel like we have so many of these discussions, and every time I have these discussions with people, everyone's like "we need to be doing something, we feel helpless, we need to be doing something," but it's so repetitive, and I think it's really great and really important to have these discussions; I think they should probably happen more frequently; it's just, it's hard because they make you *feel* so helpless.

Joey Yeah, it's like the more you talk about it, the more guilt you feel that you're just talking about it but not doing anything.

Social movement theorists tell us that one of the deeply pleasurable aspects of social movement participation is the satisfaction that comes from publicly expressing privately held values. Yet here at the other end of the spectrum we see the internal agony of those who combine awareness with a lack of sense of what can be done. It is this quality of angst, a condition that social psychologists tell us we are profoundly motivated to avoid, that essentially makes climate change "unthinkable." As Shierry Nicholsen puts it, "How many of us can really imagine that the war against nature will be over and we will come out alive in a world where continuing ecological destruction is not the order of the day? . . . We need to numb ourselves simply to get through the day" (2002, 148).

Navigating the Unthinkable: Constructing Innocence and Safety in the United States

> I find that what makes me feel guilty—it becomes this vicious cycle—I know that there are all these problems, and on the one hand I know that there are little things that I can do to alleviate those problems, so I make a conscious effort to do those, but then there are all these other problems that I know about, but to some extent I have to selectively ignore them because otherwise I feel, like, the guilt could be *so* overwhelming. So you have to focus in on specific things so that you don't get lost in that feeling of helplessness.
> —Barbara

Data from Bygdaby tell a story about how the desire to manage troubling emotions becomes the driver for ignoring climate change and how cultural tools—including norms of space, time, emotion, and conversation—together with discourses are used as mechanisms for successfully achieving that ignoring. To ignore something is essentially to manage emotion. Thinking, feeling, and talking are linked and socially structured; therefore, ignoring is also a social process. Although Americans do not have the rich, homogenous sense of tradition upon which Norwegians draw in the face of outside threats, the impulse to reach for something that brings security in the face of rapid change or of threats to ontological security or that restores a sense that one is a good person can be observed in the United States as well.

As Barbara articulated earlier, emotions of guilt, fear, and helplessness are related through a "vicious cycle" in which guilt and fear are potentially so strong that unless they can be managed, they will engender the third emotion of helplessness. The need to manage these three feelings is thus interconnected. Barbara described how she actively screens out some problems so as not to be overwhelmed by the guilt. Just as is done in Norway, one way of avoiding guilt and remaining optimistic is to control one's exposure to information. American students shared specific examples of how they and others tune out and withdraw from conversations or change the subject as methods of dealing with uncomfortable emotions. Lauren described how "people prefer to deny or ignore problems because that allows them to simultaneously deny or ignore responsibility. Also people often feel helpless by their inability to change the problem and may not want to be reminded of it." To this Robin added, "I think they're trying to avoid feeling hopeless about a problem that they feel powerless to solve." Alex indicated that he avoided people who speak about environmental problems. "If I feel I really can't do much about it, I try to avoid it." Megan described how she knows "people who avoid others who talk about climate change. They don't do this explicitly, I don't think. But I know people who 'get bored' of others who are constantly bringing up climate change because it is 'a bummer.' I think they are avoiding the other people [because they] just don't want to hear about it anymore even if they have similar opinions because it is overwhelming, a burden to always carry."

Changing the topic in conversations is one of the observations made of the focus groups conducted by the American Geophysical Union. The authors note of their participants that

As they thought about the problem, they seemed to run into brick walls, characterized by lack of clear knowledge, seemingly irreversible causes, and a problem with no real solution. As a result they were frustrated and eager for a solution but unsure of which way to go. The symptoms of this frustration are clear. The first is that people literally don't like to think or talk about the subject. Our respondents always seemed to want to move the topic from global warming itself to more familiar topics, such as moral deterioration, where at least they felt on firmer ground. (Immerwahr 1999)

Note that the reaction "I don't think about it because there is nothing I can do" is actually a fear of *experiencing* powerlessness because not looking at a problem doesn't increase one's power in the situation. As a strategy for emotion management, however, it seems to be in frequent use.

A second widespread strategy for managing disturbing emotion in Bygdaby is to focus on something one can do. As Lauren put it, "It's true, it seems that a lot of people don't want to hear about the fear because it's not a good thing that's happening, so it's a stigma, like we should have this positive, like active 'we can do something.' So the fear is sort of skipped over, 'yeah, we already know that'" (her response is followed by laughter from the group). Jocelyn, active herself on environmental problems, noted an explicit link between giving a positive message about what can be done and her feeling that she should protect herself and those around her from difficult emotions:

I feel like sometimes I can't let that space of fear and guilt get into my mind. I need to spread to the other people that I'm with that "we can do this," a positive message, [that] it's all human caused, so I feel like I can't give in to that side that wants to talk about how scary it is and how guilty I feel 'cause that's just like spinning your wheels in the mud. So when I'm talking to other people who especially don't—I'm trying to almost convince [them] that we can do this, [and] I don't want to bring up the negative aspects.

A third tactic for normalizing disturbing emotions in Bygdaby was to use humor. This method of making light of climate change comes up in the United States as well. As Joey put it, "I guess for me the only way to approach my friends like that is through joking. Making jokes about the weather or about polar bears." "Why is that?" I asked. "I think that for one they're not as seeped in the information as we are," he replied, "so they don't understand the significance, and number two, I'm sure that they also don't want to get depressed and, you know, feel guilty and feel sad and feel fearful, so I'm sure there is some cognitive dimension there, too." Ted agreed: "I was 'one of your friends' in high school. I didn't want to talk about it, I didn't want to listen. The only time we'd talk about it was when we'd have six feet of snow in April and we'd be, like, 'This isn't natural, so, sweet, six feet of snow, let's go skiing.'" Making jokes about climate change is also a way skeptics delegitimate the prospect of climate change, especially when the weather is unusually cool. Our local paper is peppered with such examples. When a large storm in January was reported in the local paper, one reader posted the ironic comment: "Global warming?" in the comment section where readers can respond to articles in the online version.

All these examples, brief as they are, illustrate the broader application of many of the techniques for normalizing in use in Norway. We can also think about parallel ways that collective constructions of time and space in the United States work to create a sense of reality in which climate change is more or less visible in daily life. Many scholars have observed that even though a large majority of Americans believe global warming is occurring and is a serious problem, a sense of urgency is lacking, in large part because many Americans believe that climate change will happen to someone else in some far-off place. Although it is true that places with more visible climate impacts, such as Bygdaby in Norway or the disappearing island of Shishmaref in Alaska, are far from "my home town," Americans' sense that such places are far away is reproduced by cultural phenomena such as the small percentage of them who are fluent in a second language, the amount of U.S. news media devoted to events outside the country's borders, and the infamously poor American grasp of global geography. The sense that climate change "will not happen to us" is manifested and reproduced by the failure of planning in relevant agencies at all levels of U.S. government. In my own community, for example, there is an almost complete lack of planning for future climate scenarios. How will changing climatic conditions affect water and power supplies? Asthma rates? The newly booming wine economy? Salmon recovery? Even after the dramatic illustration of the need to protect coastal communities from increasing storm intensity that came with Hurricane Katrina, such activities are only beginning on the federal level.

Another factor in navigating these troubling emotions concerns the extent to which they can be publicly expressed. Emotion norms are highly varied in the United States, but my environmental studies students echoed emotion factors found in Bygdaby, such as the need to maintain a sense of control. Derrick wrote that "our society has cultured us, and especially the male populace, that showing any sort of emotion or feeling is a sign of weakness and that weakness is bad. We are taught that to survive we need to keep it to ourselves, remain stoic and 'stable.' I can remember vividly in high school how I kept everything to myself, my thoughts, feelings, and fears, because I didn't want others to think I was unstable and not a 'man.'"

Norms of conversation are also part of the cultural terrain that must be navigated. In Bygdaby, one important tool for the maintenance of order and the construction of innocence comes from upholding "socially appropriate" conversation. What about conversation norms in the United States? Although the context is different, many normative patterns from Norway have parallels in the United States. Educators balance a desire to communicate information with the desire not to scare their students; local politicians consider climate change to be outside their purview; and in celebrations or social situations people don't want to bring up negative feelings. People describe awkwardness in talking about climate change. Many of my students indicated that avoidance of climate change as a topic of conversation is part of maintaining social tact. In class, students described feeling overwhelmed, but also how other social settings such as parties are not appropriate for negative topics. One student indicated that she spoke about climate change mostly in relation to class readings, discussions, or campus speakers. Climate change came up as a topic much less frequently outside of the classroom. Another student noted that she didn't talk about climate change "when I am with a large group of school friends because they think 'its boring,' especially when they are having fun. I come across as a nerd, an 'I care for the world' idealist."

One significant dimension unique to conversations in the United States is the dynamic of skepticism regarding global warming. This skepticism, combined with an American political culture in which controversy is to be avoided, can also put a damper on conversation. Unlike contexts where political debate is welcomed, in many parts of the United States expressing political views that are not shared is nothing short of bad manners. Political discussions therefore make people uncomfortable. I asked my students about this conundrum, and here are a couple of their responses:

Jocelyn I don't like conflict, and I think a lot of people don't like conflict, but they prefer not to get into that with friends, but it's not like political debates have to be personal, conflicting in that way, but that's what it's become to us. We can't have public debates about our opinions, that can be different, and we're okay with that.

Brett If there is any sort of controversial topic, it's debated over, like, the listserv. People don't have the guts to come talk to people face to face; there's no, like, public forum, no interaction.

It would seem that at least some norms of emotion, attention, and conversation in the United States muffle opportunities for ideas and serious discussion. Julie described how "even at our college where I feel like people here are really opinionated and they like to talk about things ... there is still this occurrence where when you come up on a controversial issue or people start getting riled up about something, then everyone kind of just like shies away and backs off. And I've noticed that happening pretty frequently." As a result, just as in Bygdaby, conversational spaces are constrained.

Again, norms of conversation, emotion, and attention assist people in keeping troubling emotions at bay and simultaneously produce a "double reality." They are a means through which people craft that sense of distance from climate change, that sense that this problem is not part of everyday life. These kinds of deeply embedded cultural mechanisms work so well because they feel invisible. We are rarely aware of carrying them out. Just as in Norway, however, more overt practices are an important part of socially organized denial as well. There are times and places where American complicity in climate change must be addressed and justified head on. I turn next to a sampling of such narratives.

Literal Denial, American Exceptionalism, and Other Diversions

Americans have their own set of discourses that justify climate and climate-related foreign policy, discourses that tell a story about who Americans are at the same time as they defend such policies. Official discourses in the United States look quite different from those in Norway. Whereas Norwegian discourses follow a pattern of interpretive denial (conveying a subtext that "what you see is not what you think you see"), American discourses are of the more literal "it's not happening" variety. A literal denial of the presence of climate change has taken different forms over time. Under the George W. Bush administration, suppression of reports and behind-the-scenes falsification of documents were accompanied by an official approach that the problem "needed further study." As in Norway, official climate discourses in the United States emerge from both the government and key oil industry leaders and think tanks. Peter Jacques and his colleagues fully detail the origin of discourses of skepticism and their links to the activities of conservative think tanks (centered almost entirely in the United States) (see Jacques, Dunlap, and Freeman 2008; Jacques 2009). The most recent iteration of the skeptics' "it's not happening" discourse is the attack on the message carriers as "doomsdayers" and "junk scientists" who are generally "against progress." Again, skeptic discourses appear to matter in that they contribute to both literal and implicatory denial. Although some portion of the population adopts the notion that the science is not sound, these discourses' real achievement may be the way they shift conversations away from what needs to be done to whether anything needs to be done at all. These discourses corroborate implicatory denial in that they "provide cover" for those who are not acting in response to the problem. It is interesting that the skeptics' tactics work because they undermine the condition of trust under a risk society. A risk society requires a new level of trust in science for the accurate flow of information about the physical world, but in the case of skeptics' campaigns this trust is sabotaged in the service of political and in many cases economic interests.

Another dimension of U.S. discourses is the presence of American exceptionalism in the attempt to place actions beyond examination and to position "the American way of life" as above questioning. For example, in May 2001 Bush pulled out of Kyoto on the basis that it was not in the U.S. economic interest. Just as Norwegians have put forward discourses of Mythic Norway, Americans play on the positive global role they have played regarding democracy—for example, justifying wars in Iraq and Afghanistan with discourses about spreading democracy.

Learning from the Climate Ostrich: Take Home Messages from Apathy's Ground Zero

Denial looks different in the United States than it does in Norway. The United States is obviously a much larger and much less homogenous society. Also clearly different are some of the particular reasons for ignoring climate change (such as the fact that our greater carbon emissions make our involvement harder to face) and the extent of social structural support for us to ignore it. Denial in the United States also looks different due to the presence of corporate-funded campaigns of skepticism and the increasingly successful cultural challenge they have launched to the legitimacy of science in the public sphere (see discussions on these campaigns in Freudenburg, Gramling, and Davidson 2008; Jacoby 2008; Jacques, Dunlap, and Freeman 2008; Jacques 2009).

Another important difference in the United States is that skepticism about climate change takes place in a broader cultural context of anti-intellectualism that has deep roots in American political culture and has gained a pronounced momentum in recent years (Jacoby 2008). We Americans are unique not only in our perception that climate science is contested and in our ignorance regarding the origin of climate gases (as described in numerous polls; see, e.g., Nisbet and Myers 2007), but also in our tolerance of the violations of scientific protocol and falsifications of data that have caused dozens of career scientists high up in the executive administration to resign, not to mention any number of other outcomes such as the belief that evolution is a contested theory (see, e.g., Mooney 2005; Shulman 2006). This challenge to the place of science as the basis of collective information and legitimate epistemology in the public sphere is part of the unique and changing political and cultural landscape of the United States.

In many respects and for large segments of the population, however, denial in the United States looks and feels very much the same as it does in the small community of Bygdaby in Norway. Many of the hopes and fears articulated by people in each country are remarkably similar. The widespread nature of the missing response to climate change itself indicates that there are cross-cutting, universal elements to people's experience of climate science. It is generally true, in the United States as well as in Norway and in many countries around the world, that despite new information about the fragility of our planet, life goes on "as normal." Across many sectors of U.S. society, people know facts about climate change that they believe to be true, but they live their lives without integrating this information into their decision making, political activities, or sense of daily reality. Thus, there appear to be lessons for all of us to learn from this one community in Norway. Life is different there, but for many of us it is also very much the same.

One lesson from the American context concerns the hazard of individualism. Empirical evidence from the American response to climate change also applies to ongoing discussions within political theory about globalization and democracy. It would seem that a combination of lack of trust in their political system (political alienation) and a culture of individualism leave Americans uniquely at a loss in terms of what to do about climate change. Both individualism and a declining trust that government institutions can respond to modern problems are global trends. In tandem with the expansion of neoliberal economic practice, the culture of individualism has become widespread in Western democracies around the world, especially since the 1970s. Ulrich Beck (1992, 2010) and Anthony Giddens (1991, 2009) describe it as a defining feature of postmodern society. That individualism has expanded is well established, but the significance of this expansion for human behavior, community, moral politics, and democracy is the subject of much debate. Beck (1992) points to the new possibilities for human behavior that emerge as prior social constraints fall away. As individualization frees agents from structural restraints, Beck argues, there are new potentials to engage with and effect change to the prevailing social structure. In contrast, however, Kent argues that "persistent expressions of individuals' subjection on climate change (drawn from the psycho-social evidence) are inconsistent with Beck's supposition that actors are freed through the conditions of the risk society to construct their own life courses, as this presupposes that actors possess the authority within their life realms that allow[s] them to influence and overcome the prevailing cultural and structural conditions" (2009, 144).

If individualism has reached its zenith in the United States, it would seem that one lesson American voices highlight is how powerfully this individualism further disempowers people in the face of climate change. When people are confronted with this sense of nowhere to turn, the tendency for cognitive dissonance at the level of individual psychology and collective cultural practices of denial is likely particularly strong. Under these conditions, an ambivalence regarding personal action is created, where people "choose not to choose" because they feel disempowered and ineffective in the face of the global climate challenge (Macnaghten 2003, 77). A type of "psycho-social dislocation" (Räthzel and Uzzell 2009, 333) forms under such circumstances that are rooted in the dichotomies between the individual and the social and between the local and the global.

Conclusion

> The degree of numbing of everyday life necessary for individual comfort is at odds with the degree of tension, or even anxiety that must accompany the awareness necessary for collective survival.
> —Robert Lifton, *Indefensible Weapons*

The view from Bygdaby has portrayed global warming as an issue about which people care and have considerable information, but one about which they don't really want to know and in some sense don't know *how* to know. We have watched how community members collectively hold information about global warming at arm's length by participating in cultural norms of attention, emotion, and conversation and by using a series of cultural narratives to deflect disturbing information and normalize a particular version of reality in which "everything is fine." As such, public nonresponse to global warming is *produced* through cultural practices of everyday life.

The story of Bygdaby is, on one level, a story about our collective response to climate change. But on another level it is about how socially organized denial reframes a number of existing narratives about culture, emotion, civil society, and even democracy. Through a framework of socially organized denial, our view shifts from one in which *understanding* of climate change and *caring* about ecological conditions and our human neighbors are in short supply to one whereby these qualities are acutely present but actively muted in order to protect individual identity and sense of empowerment and to maintain culturally produced conceptions of reality. This reframing has implications for our conceptions of individual caring and intelligence, for social cognition, for the role of culture in reproducing social relations, and for the centrality of emotions in culture, cognition, and political action. The social organization of denial also raises questions about new limitations for democratic process in a world where the reorganization of time and space mask the consequences of human actions. In contrast to emphases on power solely in the form of overt coercion, the notion of socially organized denial places culture and emotion at the center of social reproduction and legitimation. If we are concerned about climate change, the notion of socially organized denial clearly has implications for what to do next.

A Kaleidoscope of Conversations

How does a wealthy industrialized nation respond to the global environmental problem of climate change? Why are people not responding? This research began as a contribution to a particular thread of conversation within sociology. But in crafting this map of Bygdaby, I have drawn upon other conversations, too, and this story turns the conversation back to them as well. As I was working on the project, I came to understand that I was writing about many things at once. Watching the community of Bygdaby through first one theoretical lens and then another, and experiencing the consequent changes in perspective, I was reminded of a kaleidoscope. The kaleidoscope works as a metaphor for the varied implications of the notion of socially organized denial. If we look at Bygdaby from one view (my original one), we see a story about why people are not responding to climate change. With more time in the field, what I saw became a story about how people actively normalize climate change. Thus, from a more theoretical standpoint, we have a story about emotions and culture, how emotions and emotion management shape conversations, awareness, and sociological and political imagination. The story from Bygdaby is also a story about how an apparently politically and economically powerful people, people who seem to "have it all," are paralyzed in the face of not only climate change, but the many problems of the modern world, from human suffering in the form of disease, warfare, and human rights abuse to ecological deterioration. As such, it is a story about the precariousness of privilege, the darker consequences of the good life in a world where not all have that life. The problem of socially organized denial tells a story about links between macroeconomic structure, community sensibilities, and individual emotions. The story of Bygdaby thus describes how economic and political structures can permeate our most private moments and minds.

Environmental Sociology, Social Psychology, and Science Communication

Our first view through the kaleidoscope takes us to a conversation about public response to climate change. The conversation is populated with scholars from a series of disciplines, all of whom "get" the serious consequences of climate change. These scholars wonder with some urgency why the public fails to appear concerned. They wear the hats of environmental sociology, social psychology, and science communication (see, e.g., important work in Dunlap 1998; Ungar 2003, 2007, 2008; Moser and Dilling 2004; Brewer 2005; Weiskel 2005; Leiserowitz 2006, 2007; Moser and Dilling 2007; Moser 2007b; Dilling and Moser forthcoming). Most of these scholars are social scientists, but some are biologists and climate scientists who work on interdisciplinary teams. They also wonder, with no small measure of urgency, what can be done to get the public on board for the political and social changes required to bring carbon emissions to sustainable levels. The story from Bygdaby contributes to their conversation. My research was originally conducted from the viewpoint in the room that these scholars occupy and was intended to contribute to their discussion.

I began this project with the sense that existing studies of how people respond to climate change, although clearly valuable, had paid too little attention to the role of political economy on the one hand and to social psychology and culture on the other. Working from the ground up allows us to include these important dimensions and thus bridge the troublesome yet nevertheless ultimately artificial gap between micro- and macro-social processes. Most existing studies on public response to climate change—coming from environmental sociology, social psychology, or science communication, from survey work on attitudes and beliefs to psychological studies of mental models—use individuals as their unit of analysis. Yet as we have seen, social context itself can be a significant part of what makes it difficult to respond to climate change. Although findings from this literature are essential, studies of perception that focus solely on individuals are unable to grasp the meaning of differences across cultures, subcultures, nationality, or the influence of political economic context on how individuals and communities think, feel, and imagine.

This view from the ground up in Bygdaby adds to an existing understanding of public responses to climate change within environmental sociology in two important ways. First, my work brings to environmental sociology a deeper emphasis on the feelings that people have about climate change and the ways in which these feelings shape social outcomes. Although not normally applied to environmental issues, research on the sociology of emotions is highly relevant to understanding community members' reactions to global warming. Emotions form the texture of how we embody society: "our emotions form part of our point of view of the world; we do not just have them, we exist in and by way of them" (Crossley 1997, 27–28). In this case, emotions are key in shaping social response. Second, the view from Bygdaby brings to our attention relationships among individual responses, culture, and political economy. Individualistic approaches to the question of climate apathy that constitute the bulk of such studies obscure the existence of "social facts," the way in which individuals' seemingly rational actions are in fact merely reflections of permissible patterns of behavior within a particular social structure. There is a long tradition within sociology of tracing relationships between social context and one's view of the world. Recall how Émile Durkheim, in 1915 described how social existence determines social consciousness—that is, that the most basic categories of thought, our "cognitive architecture," derive from the social conditions of existence. Marx's concept of ideology and Dorothy Smith's work on feminist standpoint theory highlight how mental structure is determined by social structure. In contrast to a focus on individuals, the view from Bygdaby examines public response to climate change in the context of a *contradiction* between environmental values and political economy, between knowledge and everyday practice. Voices from Bygdaby show that it is by paying simultaneous attention to individual responses and social context that we can begin to look at relationships between people's reactions to global warming and the larger political economic system. Cognition, awareness, and denial emerge from a process that spans micro-, meso-, and macrolevels of social structure. At the microlevel, we see the role of individual emotions and interactions. At the mesolevel of culture, norms of behavior from conversation, feelings, and attention shape and reinscribe what is considered "normal" to think about, talk about, and feel. And these local cultural norms are in turn connected to larger political economic relations. Furthermore, through the use of cultural tools, individuals are able to manage troubling emotions. Thus, acts of emotion management are in turn situated within and work to reproduce mesolevel cultural resources for emotion management, which in turn normalize national macrolevel economic and political policies (e.g., high carbon emissions, the expansion of oil and gas production). The causal arrow goes both ways. Macrolevel polices are part of what troubles individuals (e.g., generating emotions of guilt), but at the same time the national government provides narratives that individuals use as justifications for behaviors.

When social context is taken into account in terms of both local cultural norms and larger political economy, two significant contradictions emerge. First, there is the contradiction between privately voiced feelings and the larger cultural norms of emotion, conversation, and attention mentioned earlier. That is, individual feelings of concern, powerlessness, and guilt are experienced in the context of social pressure to fit in, an absence of spaces for emotional or conversational expression, and specific emotion norms of toughness and maintaining control. Second, both emotions and knowledge are experienced in the context of a contradiction among national values, history, and identity in connection to environmentalism and political economic relations that make present-day behaviors with respect to global warming antienvironmental.

Social Organization of Denial

Now we can again turn the kaleidoscope and peer into a second conversation. The concept of denial is generally considered the domain of psychology. But if social existence determines social consciousness, then the information individuals find disturbing and the mechanisms they employ to protect themselves from such information must also be analyzed within the context of both social interaction and the broader political economy. Sociologists remind us that notions of what is normal to think and talk about are not given, but instead are socially structured. In Bygdaby, multiple aspects of social context from high carbon emissions and oil wealth to emotion norms of maintaining control are a significant part of what makes it difficult to respond to or even think about climate change.

I am not the first sociologist concerned with the topic of denial. I have drawn here upon the work of Stanley Cohen, whose book *States of Denial: Knowing about Atrocities and Suffering* (2001) lays out important concepts regarding denial as a collective social process, including the typologies of literal, interpretative, and implicatory denial. As his title implies, Cohen is particularly focused on human rights abuses, but his work points to the role of social structure in the processes he describes.

Building on Cohen's typologies, I have provided specific examples of a range of denial tactics in use in the community of Bygdaby and demonstrated how everyday life practices can reproduce denial. My analysis leans especially heavily on Eviatar Zerubavel's (1997, 2002, 2006) work on the social organization of denial. His studies highlight the role of social cognition in the reproduction of power in ways that are particularly relevant for the flow of information on climate change. He reminds us that "through the various norms of focusing we internalize as part of our 'optical' socialization, society essentially controls which thoughts even 'cross' our minds" (1997, 51). Although much attention has focused on the important issue of availability of information and the limited access Americans have to information as a result of corporate-controlled media (Chomsky 2002; Herman and Chomsky 2002), Zerubavel emphasizes the complementary relationship between secrecy and silence. Media sources not only provide minimal information, but also teach the public what to notice and what not to notice and tell people (or fail to tell them) how information on climate change is connected to their daily lives. Although there is information about climate change that remains secret from the general public, there is more than enough information available for anyone to realize that the problem is significant. In other words, Zerubavel's analysis prompts us to notice how much information is available but goes unnoticed due to systems of perception and attention, how people forget this information when they do hear it (due to socially shaped systems of memory), how information is not considered morally offensive (being outside socially defined limits of concern), and finally how information on climate change is not connected to other environmental problems or to personal life (due to socially shaped systems of organization). Secrets help solidify structures of denial and ignorance, but Zerubavel emphasizes that blocking information from our own awareness and preventing it from entering others' awareness are functions that fit together. It is because of the combination of these socially shaped systems of thought, memory, and cognitive organization that denial feels so much like everyday life.

This existing work on the social organization of denial is enriched by the view from Bygdaby in several ways. In blending material from political economy, social psychology, emotions, and culture in an ethnographic case study, this project clarifies why it makes sense to conceive of denial as a socially organized rather than merely individual phenomenon. If Zerubavel reminds us that individual acts of avoidance are socially structured, I expand this assertion by illustrating how what individuals choose to pay attention to or avoid reflects not only social norms of attention, but also norms of emotion and conversation. Building further on other cultural theory, I highlight the production of thought patterns through the fact that individuals are motivated to manage their emotions and do so via a shared tool kit of resources for distancing themselves from disturbing information. I have also linked the study of denial with a number of other key issues, including privilege and environmental justice. Finally, I have added a number of what I hope will be useful concepts to the field, including "tools of order," "tools of innocence," "moral imagination," and "the social construction of innocence."

Sociology of Emotions

The in-depth examination of one setting—the case study approach—allows us to make connections between diverse theoretical frameworks. Bringing denial, the notion of a cultural tool kit, and questions from environmental sociology to the field of the sociology of emotions turns the kaleidoscope as well and thus creates new patterns. Perhaps the most central questions within the sociology of emotions concern the roles of emotions in maintaining social order and leading to social change. Like conversations, emotions matter in the generation of a public sphere. Emotions may unite people in common cause, inform their interpretations of the world, and thus catalyze both the sociological imagination and political power. But not inevitably. Here the view from Bygdaby provides some insight. Emotions can be a source of information (Hochschild 1983) and an impetus for social action (Jasper 1997, 1998; Polletta 1998), but my observations in Bygdaby highlight how the desire to avoid unpleasant emotions and the activities of emotion management can also work against social movement participation. Emotions play a key role in denial, providing much of the reason why people prefer not to think about global warming. Furthermore, the story of Bygdaby is largely about how emotion management is central to the *process* of denial. Finally, we see links between theory on emotions and culture in that the management of unpleasant and "unacceptable" emotions is carried out through the use of a cultural stock of strategies and social narratives that are employed to achieve selective attention, perspectival selectivity, and the stopping of thought. Thus, movement nonparticipation in response to the issue of global warming in 2000–2001 did not simply happen in Bygdaby but was actively produced as community members kept the issue of global warming at a distance via a cultural tool kit of emotion management techniques.

Although this project has drawn most heavily on theory from the sociology of emotions, it is worth noting that there is a parallel literature in psychology that deals with "affect" and cognition as well as a literature on science communication that applies psychological concepts and work on emotions to climate change. Social psychologists describe how *affect*, the positive or negative evaluation of an object, idea, or mental image, has been shown to powerfully influence individual processing of information and decision making. Anthony Leiserowitz (2006, 2007) describes how climate change evokes negative affect. Work in the area of risk perception and affect in the United States and Great Britain by Irene Lorenzoni and her colleagues (2006) similarly found that respondents reacted negatively to any terms having to do with global warming or climate change.

Valuable research on climate change has recently included emotions (see, e.g., Stoll-Kleemann, O'Riordan, and Jaeger 2001; Meijnders, Midden, and Wilke 2001b; Swim, Clayton, Doherty, et al. 2009). For example, Susanne Stoll-Kleemann, Tim O'Riordan, and Carlo Jaeger (2001) describe how their Swiss focus group participants experienced fears about future climate scenarios and dissonance with respect to their contribution to the problem and how they developed justifications for their own inaction. Furthermore, the authors note how "to overcome the dissonance created in their minds [the participants] created a number of socio-psychological denial mechanisms. Such mechanisms heightened the costs of shifting away from comfortable lifestyles, set blame on the inaction of others, including governments, and emphasized doubts regarding the immediacy of personal action when the effects of climate change seemed uncertain and far away" (107). In contrast to work on how emotions can mute social change, Anneloes Meijnders, Cees Midden, and Henk Wilke (2001a, 2001b) describe the positive role of fear in information processing related to energy conservation. Again missing from these accounts, however, is the link between emotions and social structure, either in terms of how social structure and political economy shape emotional experiences or in terms of how emotions become part of the reproduction of power (through legitimation and denial) or the reorganization of power in the case of the social movement process. Emotions matter for how and whether groups of people will develop a social or political imagination or acquiesce to an existing ideology. In addition, although psychological explanations such as cognitive dissonance have provided powerful insight into why emotions are a *cause* of denial, sociological accounts further tell us more about emotions in the *process* of denial.

Tools of Order and Innocence

If we turn the kaleidoscope yet again, we can see the story of Bygdaby in relation to the sociology of culture. Questions about how people produce everyday reality and how they normalize troubling information presume both a level of agency among individual actors and the everyday life world's significance to political outcomes. In both these realms (individual agency and the everyday life world or lived experiences), there are relevant conversations within the sociology of culture. Using Anne Swidler's (1986) model of culture as a tool kit and Jocelyn Hollander and Hava Gordon's (2006) thirteen tools of social construction, I have described how people create a sense of everyday reality using features of everyday life from emotion norms to cultural narratives. Thus, denial is "socially organized" not only because we think as members of groups, but because societies develop and reinforce a whole repertoire of techniques or "tools" for ignoring disturbing problems. Norwegian emphasis on tradition affirms a sense of order, and public discourse about the superiority of Norwegian natural gas (which supposedly produces lower carbon dioxide emissions) or high levels of humanitarian aid affirm a sense of Norwegian innocence. Two sets of tools are utilized to ignore problems: "tools of order" affirm a sense of how things are in the world, their ontological stability, and "tools of innocence" create distance from responsibility and assert rightness or goodness of actions. Throughout this story, culture is clearly itself a realm of social action, but it is also tied to the microlevel of emotions and the macrolevel of political economy.

My sense of the importance of conversation leans on the important work of Nina Eliasoph, who is concerned with political silence in the United States on a range of issues. Her book *Avoiding Politics: How Americans Produce Apathy in Everyday Life* (1998) emphasizes the role of conversation and conversation norms in how people create cultures of political silence. Eliasoph draws our attention to conversation as a means of producing (or diminishing) the public sphere. Just as there are political implications to conversation in terms of the development of awareness and construction of reality, control of emotion also has deeply political implications. It is through a combination of emotion management and conversations that people collectively produce apathy.

Privilege, Invisibility, and Environmental Justice

How does collectively ignoring the problem of climate change serve to maintain Norwegians' position of relative economic privilege? Robert Lifton (1967) coined the term *psychic numbing* and writes articulately on the topic. Yet his work does not explicitly direct our attention to the fact that in the process of numbing there is also the maintenance of privilege. Social psychological accounts of identity or cognitive dissonance do not make such connections, either. Understanding these links among denial, cognitive dissonance, and privilege affords us another important view through the kaleidoscope. Situating cognitive dissonance and denial in the context of privilege underscores the relationships among individual emotions, culture, and social structure and allows for the incorporation of potent concepts from the study of privilege (see, e.g., Kimmel and Ferber 2003). When combined, all these insights allow us to see the powerful effect that structures of invisibility have in the silence surrounding climate change and the perpetuation of global environmental degradation.

Most work on privilege examines gender, race, and class privilege, and most environmental justice work "studies down" in this context: privilege and environmental justice are usually linked through descriptions of the unequal burden of the "underprivileged." Instead, my project looks at the lived experiences of those who reap the benefits of global environmental inequality. Linking denial with privilege also allow us to understand the story of Bygdaby as one that lends a human face to the experience of privileged people. This view is important. Figuring out why and how middle-class and wealthy people in the global North perpetuate environmental problems is as crucial to the field of environmental justice as critical white studies is to the field of race or masculinity studies to the field of gender. For people of color living on low-lying islands or struggling from flooding or drought, the key questions relating to climate change may be how effectively to bring attention to their plight and justice to their lives. But for middle-class white environmentalists like myself living in wealthy nations, the key questions related to climate change have to do with denial: Why are so many people in the first world so willing to live in denial? How is this denial managed? What does it look and feel like? What are its personal and social consequences? Are there any weak points in the armor of this denial?

When I was in my early twenties, I was captivated by a paradox: How could it be that almost all the individual people I met seemed intelligent, caring, and basically decent, but at the same time such profound suffering, indifference, and exploitation continue in the world? How can "good" individual people collectively generate such "bad" social and ecological outcomes? This paradox eventually guided me from an undergraduate degree in biology to graduate work in sociology. I believe it is valuable to apply this same double view of privileged people to the situation of climate denial. In other words, let us approach privileged people simultaneously through the lens of compassion and the lens of critique. Privilege is a precarious position. People occupying privileged social positions encounter "invisible paradoxes"—awkward, troubling moments that they seek to avoid, pretend not to have experienced (often as a matter of social tact), and forget as quickly as possible once those moments have passed. When it comes to our relationships to environmental problems, Shierry Nicholsen reminds us,

> In the current situation of ongoing environmental destruction, those of us who live in the West (and certainly some who live elsewhere) are all in some degree both perpetrators and victims. . . . We know we are caught up in a system that causes damage and must cause further damage simply in order to continue operating. At the same time, we know that we and others are suffering from that damage, to whatever degree, and that we are all likely to suffer still more directly and more painfully in the future. On both counts, we feel guilty. How can we do more? How could we have participated in doing what we have already done? How can we live knowing that we are part of this impossible situation? (2002, 141–142)

The voices we have heard from Bygdaby and the United States are those of people who are deeply troubled, who very much want another world, who would do a great deal to create change if they only knew how, who feel they can make a difference. Recall that cognitive dissonance is generated by threats to identity and individuals' desire to preserve a positive sense of self-esteem. Cognitions related to climate change are avoided because they create such dissonance and are thus troubling. Ironically, if they were less troubling, perhaps more social engagement would be possible. Voices from Bygdaby and the United States tell a story about how profoundly difficult it can be to be both aware and informed at this point in human history.

But we also need to be critical of the collective outcome when these individually "good" people knowingly emit enormous levels of carbon dioxide in order to travel for their tropical vacations, fail to call their leaders to account, and justify their actions through normalizing narratives. Antonio Gramsci went to prison with the conviction that power lies in the sphere of culture. Privileged people reproduce existing power relations as they enact denial in everyday life. It is only by using this double view of compassion and critique that we are pointed to the deeper challenges to individuals, and threats to the democratic system posed by climate change. If all we need to do is better educate or better socialize the population as a whole to the moral consequences of their actions, we can say that our social system is basically fine. It might require only a little tinkering. It is when we apply lenses of both compassion and critique, however, that we see the complexity of human nature, the complexity of our globalized society, and the real challenges to both that climate change brings forth.

Existing work on privilege highlights a number of useful concepts, especially the aspect of invisibility and the notion of innocence. The privilege of being a citizen of a nation in the wealthy first world has parallels to class privilege, especially in terms of the quality of invisibility. Michael Kimmel describes how for Americans "class privilege feels like something we have earned all by ourselves. Therefore class privilege may be the one set of privileges we are least interested in examining because they feel like they are ours by right, not by birth" (Kimmel and Ferber 2003, 8). Kimmel's observations that privilege feels rightfully earned and therefore is not a case of privilege at all are directly parallel to the logic of the "we have suffered" narrative I described in chapter 5 or the discourse of American exceptionalism. It is because of this invisibility that people in privileged positions can and do easily ask, "What does this problem have to do with me?" Although *Bygdabyingar's* higher levels of education and value orientations have given them a greater sensitivity to what climate change has to do with them, there still remains the gap between daily life, which feels like it has nothing to do with oil production or climate change, and this very significant global environmental problem. At the scholarly level, we can also see that privileged people's experiences and normalizing strategies have remained invisible in that they are regarded as neutral and universal within social psychological theories (which also largely fail to incorporate inequality or power).

The concept of denial in turn brings important emphases to existing work on privilege. This new linkage between denial, privilege, and environmental justice draws our attention to the concept of "environmental privilege" emerging from the work of Laura Pulido (2000), William Freudenburg (2005), and Raoul Lievanos (2010). Lievanos describes environmental privilege as "the taken-for-granted structures, practices, and ideologies that give a social group [a] disproportionately high level of access to environmental benefits" (2010, 1). "Environmental privilege" in this case applies to the privilege of living in a "core" or advanced capitalist nation, the privilege of being a producer of oil and natural gas, the benefits of which are garnered at home and the hazards of which are exported into the common airshed. Although study of climate change and the social impacts of other environmental problems has taken place mostly within environmental sociology, climate change exacerbates existing social inequality within countries and across the globe. Integration of work on climate change, migration, human rights, and human security is taking place in the policy realm, but there is still too little crossover within academic scholarship.

It is when we incorporate the concepts of privilege and denial that we ultimately come to see why social inequality and the reorganization of time and space under globalization so forcefully perpetuate environmental degradation. Anthony Giddens describes this new organization as "time-space distanciation": "globalization can thus be defined as the intensification of worldwide social relations which link distant localities in such a way that local happenings are shaped by events occurring many miles away and vice versa. . . . In the modern era, the level of time-space distanciation is much higher than in any previous period, and the relations between local and distant social forms and events become correspondingly 'stretched'" (1991, 64).

Distanciation reshapes the landscape of privilege and produces new possibilities for denial. Through both distanciation and privileged actors' active production of a safe, secure mental world, environmental problems are kept invisible to those with the time, energy, cultural capital, and political clout to generate moral outrage in a variety of ways. Wastes and hazards are exported to other nations, other places, other populations. Privileged people are protected from full knowledge of environmental (and many other social) problems by national borders, gated communities, segregated neighborhoods, and their own fine-tuned yet unconscious practices of not noticing, looking the other way, and normalizing the disturbing information they constantly come across. As a result of the forces of distanciation and denial, the environmental problem of toxic waste is invisible to those who do not live near hazardous sites or who can move, hire lawyers, and effectively make a fuss if they do. The environmental problem of contaminated water feels invisible to those who can easily afford to buy their water bottled. And the issue of climate change will affect nations with less infrastructure long before it will touch the lives of Norwegians, even if most Norwegians are avid skiers. Ecological collapse comes to be seen as a fanciful issue to those of us in the "safe" and "stable" societies of the North as we buy our fruits and vegetables from South America and our furniture from Southeast Asia, but also as we send our wastes into the common atmosphere.

Moral Imagination or Construction of Innocence?

Everything is coming closer and faster: the faces of people in agony, the space and time it takes to reach them, the life-saving work of doctors or engineers. The boundaries of "moral impingement" have been widened.
—Stanley Cohen, *States of Denial*

Any nation who believes that it began innocent has a sentimental mentality. We don't know of a nation in the modern world that is not linked to some form of barbarism, some structures of domination shot through its past and its present. There are degrees and gradations, yes, but the very notion of being born innocent is a sign of what Henry James called a "hotel civilization." A hotel civilization is obsessed with comfort, convenience, contentment, where the lights are always on. It's predicated on the notion that the sun is always out, thereby denying your night side and your dark side.
—Cornel West, from "America's Night Side" speech at the International Press Institute World Congress, Boston

The questions and contradictions raised by the issue of climate change apply to a variety of present problems and are relevant for wealthy people around the world. We live in a world that is increasingly complex. With the information age, the experiences and perceptions of space and time for people in places such as Bygdaby have been dramatically reorganized in the past generation. People in wealthier parts of the world are now learning a great deal about "distant" places. At the same time, more of their food and clothing are imported from poor nations of the world, and more of their waste products are exported there. These ongoing global changes in social organization create a situation in which, for privileged people, environmental and social justice problems are increasingly distant in time or space, or both. Global capitalism is currently producing wider divides between the material conditions of the lives of haves and have-nots, and closing gaps in space and time through information technologies, bringing privileged people ever closer to at least some aspects of the worlds of those they exploit through cheap airfares and quality digital Internet images. Cohen describes how

> the free market of late capitalism—by definition a system that denies its immorality—generates its own cultures of denial. More people are made superfluous and marginal: the deskilled, unskilled and sinking poor; the old, who no longer work; the young who cannot find work; the massive shifting populations of migrants, asylum seekers and refugees. The "solution" to these problems now physically reproduces the conditions of denial. The strategy is exclusion and segregation: enclaves of losers and redundant populations, living in the modern version of ghettos, remote enough to become "out of sight, out of mind," separated from enclaves of winners, in their guarded shopping malls, gated communities and retirement villages. (2001, 293)

If our species is going to change the course of global climate change, it is important to understand denial of that change from a strategic sense. But, as if this task were not enough, denial is also important to understand because it provides a window into a wholly new and profound aspect of the experience of modern life. It is an outcome of a world where for millions of people a keen moral socialization and a belief in equality collide with more information about the vast inequality of economics and life chances of people than ever before. Climate denial is an outcome of a world in which time and space have been restructured such that the most intimate details of life from food, clothing, or family vacations are directly yet invisibly linked to the hardships and poverty of people in other parts of the world. Climate denial is a consequence of a world in which boundaries that once existed are collapsing. It is a world where a sense of security and order—if they are to be achieved—must be reasserted through cultural practices as well as through physical walls. Widening gaps in global income and increases in international trade resulting from the globalization of capitalism lead to the displacement of environmental and social problems across national boundaries, onto other people (and species), and across time (through extinction, sea-level rise, and future weather scenarios) and space (through melting polar ice caps, floods, food scarcity, and disease in tropical nations). Under these conditions, the kinds of contradictions between values and actions that Bygdaby residents face become simultaneously more visible and further obscured. Because of the dynamics of global capitalism in which gaps between rich and poor increase, environmental problems are displaced across space and time, and information is widely available to people in wealthy nations, the issues of denial, privilege, double reality, and global environmental justice will become increasingly salient for the "educated and comfortable people living in stable societies" (Cohen 2001, xvi). There will be more and more opportunities for people both to become aware of *and* to "look innocent by not noticing" troubling information (Cohen 2001, xii). Privileged people around the world will be faced with more and more opportunities either to develop a moral imagination and imagine the reality of what is happening or to construct their own innocence from the resources of their culture's particular tool kit.

Socially organized denial in the face of climate change challenges us both as individuals to act as moral agents and to retain our sense of positive self-identity, self-efficacy, and hopefulness in the face of ever greater obstacles and as a democratic collective to develop an authentic means of collectively participating in the direction of our future. So what are we to do? If our "problem" is redefined from lack of information or caring to one of apathy as suffering, how should this redefinition change the course of our response? We must apparently go beyond information and environmental education campaigns. But how so? Everyone wants to know how we overcome denial.

Individuals: A Struggle to Imagine the Real

We are in considerable ignorance about how significant changes in human consciousness come about.
—Robert Lifton, *Indefensible Weapons*

We also know that a sense of absurdity, if acknowledged and confronted, can lead to something beyond itself. And from that very struggle to transcend absurdity can emerge the most powerful kinds of learning.
—Robert Lifton, *Indefensible Weapons*

Globalization constrains democracy due to a reorganization of our sense of citizenship, our sense of empowerment, and our sense of commitment to moral community. We fail to imagine how our lives are connected to those of the people who grow and harvest our morning coffee (despite the fact that our bodies come in contact with the same chemicals as theirs), how our values are threatened by the working conditions they face, how our lives are impacted by the exposure that we both face. C. W. Mills writes, "The very shaping of history now outpaces the ability of men to orient themselves in accordance with cherished values" (1959, 4). When nation-states are less important, we are less able to impact decisions connected to our immediate lives (such as the production of our morning coffee) because we cannot vote in Guatemalan elections or as shareholders of Starbucks. But this does not mean that our lives are not in fact connected to these other people. How do we learn to see the connections?

One implication of socially organized denial of climate change is that as individuals we must struggle to imagine the reality of our current situation. In writing on the nuclear peril, a problem that now seems infinitely more manageable than climate change, Lifton describes many of the same difficulties we face in coming to terms with climate change. He writes of our "fragmented awareness," how "we have no experience with a narrative of potential extinction," and how therefore we "cling to a desperate conventionality." He points out that the emotion of fear inhibits our ability to break through "illusions" to "awareness." And at stake in our "struggle for awareness" is the fact that "the degree of numbing of everyday life necessary for individual comfort is at odds with the degree of tension, or even anxiety that must accompany the nuclear awareness necessary for collective survival." He notes that with the appearance of nuclear weapons, imagining the reality of our situation became "uniquely difficult, and at the same time, a pre-requisite for survival" (1982, 117, 108, 5). Thinking and feeling are linked to each other and bound up with imagination. How can we imagine the reality of climate change when we have no prior understanding that connects with it? As a result, to borrow language from Lifton again, we are "haunted by something we cannot see or even imagine" (1982, 4). The question of what is to be done to break through climate denial is among other things a question about how to reinvigorate the sociological imagination in this new context where the consequences of our own actions are increasingly mystified. How do people understand who they are under the impacts of their actions and respond as moral agents under these conditions? How do they cultivate a moral sensibility?

We may not know enough about how to face this challenge, but for those who do break through and become politically engaged, it is an empowering act. As we cross this threshold of awareness, Lifton tells us, "we extract ourselves from the roles of victim and executioner, as well as immobilized bystander." He also tells us that crossing the threshold into awareness "is accompanied by exhilaration, which has to do with a new sense of integrity between self and world. In the process one may shed several forms of self-condemnation bordering on guilt, move toward responsibility" (1982, 122, 120).

In thinking about what it will take to mobilize individuals in relation to climate change, the work of Susanne Moser is particularly valuable (see Moser and Dilling 2004; Moser and Dilling 2007; Moser 2007b; Dilling and Moser forthcoming). Moser takes stock of psychological and structural barriers and advises that "information and understanding by itself typically do not suffice to motivate behavior change or civic engagement." Instead, she and her coauthor Maggie Walser recommend that "it is critical now for communicators to constructively engage and support individuals and communities by creating a sense of feasibility, collectivity, and urgency arising from fact, experience, common sense, and a moral sense of responsibility" (Moser and Walser 2008). From here, they offer specific message points, including the notion that global warming is not a future problem but a present challenge; that many people, communities, and businesses are already involved; and that people who have taken first steps have actually saved energy and money and have improved quality of life and local economies by reducing traffic congestion, improving air quality, and more.

Moser and Walser also incorporate psychological theory on people's need for self-efficacy when they direct us to the need for hope. Under the subhead "Scientific Confidence, Practical Solutions, and Hope," they emphasize the importance of communicating solutions:

What is needed more now is information about practical solutions, help, support from others, encouragement, and empowerment. What is needed now is a sense of hope. Tapping into people's desires for a better future, their social identities and aspirations, and cultural values that promote individual and collective action and engagement for the greater good (e.g., ingenuity, responsibility, stewardship, being a good team player, and leadership) can all increase people's motivation besides the more instrumental reasons. (2008)

As powerful and necessary as most climate communications strategies may be, they have a limitation. They are aimed at individuals, but individuals cannot solve climate change on their own, and they know it. Emissions reductions at the scale we need must be achieved at the collective level. For Giddens, a central part of the problem here is that "we have no politics of climate change" (2009, 4). Climate change challenges our existing social order like no prior social or ecological problem, so developing that politics is no straightforward matter. Climate change is disturbing not only because of the social and ecological collapse that is predicted, but because of the apparent possibility that our existing political structures are not up to the task. Unless we can also refashion our political and economic systems, we are trapped. We need to imagine the reality of our lives and break through the absurdity of the double life. However, we face a circular situation. The fact that many people perceive no viable political options is a central part of why they are not responding. Individual apathy is a rational response if there is nowhere to turn. In order to mobilize people, we need real solutions. And here we see a second limitation to many efforts. Because people find real change unnecessary, beyond their imagination, or politically unfeasible, they offer or look to solutions that tinker with the system instead.

In a recent debate with communications expert George Lakoff, sociologist Robert Brulle describes how

> the idea that more effective environmental messages, developed through the application of cognitive science by professional communication experts, can favorably influence public opinion, and support legislative action to remedy this issue . . . lack[s] any contextual basis within a larger theoretical structure of the role of communication. This theoretical deficit leads to the development of climate messaging strategies that support short term pragmatic actions that fit thin economic and political imperatives, but fail to address meaningfully the ecological imperatives defined by global warming. (2010, 83)

Climate change requires large-scale reduction of emissions, but our current political economic structure is intimately embedded in our petroleum-based economy. Thus, Brulle observes, "while identity campaigns may offer short-term advantages, they are most likely incapable of developing the large-scale mobilization necessary to embrace the massive social and economic changes necessary to address global warming" (2010, 83).

Climate Denial, Globalization, and Democracy

Individuals matter for democracy and vice versa. Alexis de Tocqueville wrote of the peril of "soft despotism"—that threat to democracy in which "each [individual], withdrawn into himself, is almost unaware of the fate of the rest" (1969, 692). Tocqueville wrote and worried about this threat to democracy in the 1830s. As the world has changed, the problem of soft despotism seems to have both increased and spread around the world. Robert Bellah and his colleagues have more recently written of a "crisis of civic membership" in the United States, which is "expressed in a loss of civic consciousness, of a sense of obligation to the rest of society, which leads to a secession" (2008, xviii). If climate change poses a new challenge for individuals to live morally coherent and responsible lives, it simultaneously poses a new challenge for democracy. As things now stand, the human capacity for denial in the face of problems that feel too large to tackle threatens to erode dangerously the critical democratic role of the public sphere at a time when we would seem to need it more than ever. Without sensing a genuine reason to engage, individuals withdraw from the political as a self-protective response. Yet for democracy to work, it needs engaged citizens thinking and caring and acting ethically on behalf of the larger society.

Climate change and denial of climate change are interwoven with accelerating globalization, technological change, and time–space distanciation. For both Anthony Giddens (1991, 2009) and Ulrich Beck (1992, 2010), the new circumstances of individualism and globalization enable a greater degree of individual freedom at the same time as they require of each of us a new level of self-awareness or "reflexivity." Beck points to the new and positive possibilities for human behavior that emerge as prior social constraints fall away. He argues that as individualization frees agents from structural restraints, there are new potentials to engage with and effect change upon the prevailing social structure. Globalization creates more autonomy and choice for individuals, says Beck, but carries new burdens in the form of decision making and new responsibilities.

Yet when it comes to climate change, it would seem that individuals on their own can neither garner the necessary political power to effect change nor pull off the needed quality of reflexivity that Giddens would have as the antidote. Instead, we see how individuals are paralyzed by troubling emotions and are prone to reach for the cultural tools of order and innocence that are the building blocks of denial. Jennifer Kent argues that the evidence is on the table: when it comes to climate change, individuals lack the reflexivity that would enable effective democratic response. "Empirical psychological and sociological evidence of individual attitudes and behaviors towards global climate risk conditions . . . suggest that the processes of transformative social change are yet to emerge" (2009, 141). Instead, as Michael Maniates so well describes, individualism and globalization work together to create a disjunction between "our morals and our practices" (2002, 51). The crisis of climate change makes clear how badly we need a mode of social organization that promotes organized responsibility rather than organized irresponsibility and denial. But how do we get from here to there?

Can We Make the Path by Walking It?

Caminante, son tus huellas
 el camino, y nada más;
 caminante, no hay camino,
 se hace camino al andar.

Wanderer, your footsteps are
 the road, and nothing more;
 wanderer, there is no road,
 the road is made by walking.

—Antonio Machado, "Proverbios y cantares XXIX"[1]

Climate denial leaves us with big questions: How do we break through denial into awareness? How do we reinvigorate political and economic systems? How do we move forward in the face of enormous uncertainty? With the rise of globalization, social movements around the world have emerged based on a fierce return to the local. Norwegian farmers take seriously the politics of "local patriotism" and local dialects. The Slow Food movement launched in Italy in 1986 has become a vibrant force around the globe. In the United States, we see the rise of community-supported agriculture, farm-to-school programs, and a surge of local produce in restaurants. The environmental movement in the western United States has been radically reinvigorated by a movement of collaborative conservation from the ground up (Brick, Snow, and van de Wetering 2001). Although not inherently unproblematic, such movements may provide a key for breaking through climate denial.

If we follow this lead and work from the ground up, perhaps we too can make the path by walking it. There is already a momentum building for communities to uncover how climate change is manifesting in their local contexts and from there to respond. Working from the local outward simultaneously reinvigorates people's life world. We can engage communities in projects at the local level in which they assess climate impacts and develop solutions. How might shifting weather patterns affect local agricultural crops and in turn regional economies? Will transportation routes be affected? Songbird populations? Fire hazard around homes? Climate impacts may be particularly easy to visualize in rural areas, but tangible questions about impacts and their mitigation can be applied to urban areas as well. The very participation in these kinds of planning actions will reduce the gaps between abstract information and daily life, decrease the sense of a double reality, and bring home impacts in economic, infrastructural, and physical terms. There has been debate about how much attention should be focused on "climate mitigation" versus "prevention" efforts. Yet in this case the two can go hand in hand.

Local community groups can likewise calculate the carbon footprints of homes, schools, churches, and upcoming city-planning efforts and use this information to make reductions. Once the terrain is better understood, concerned citizens and city councilors alike can ask whether there are things that communities can do to mitigate these impacts. Such efforts are already under way. Susanne Moser describes how the absence of strong climate leadership in the United States left a void that "created fertile ground for bottom-up political mobilization" where numerous civic, private, local, and state groups are now active. "The evidence suggests that momentum is quietly building in the shadows of federal inaction regarding mandatory emission reductions. The examples—merely indicative of a much larger array of actions—paint a picture of numerous sub-movements not yet merged and recognizable as one" (2007a, 140). And Carlo Aall, Kyrre Groven, and Gard Lindseth describe how in Norway local municipalities are also starting to play an important role in climate policy: "Since Norway so far does not have any well-developed national policy on involving local government in fulfilling the national GHG [greenhouse gas] emission reduction goals, all those municipalities which have developed a comprehensive local climate policy could in principle be defined as examples of municipalities playing the role of policy actors within the field of climate policy" (2007, 97).

What is important for effectiveness, however, is that people understand and experience activities as more than ends in themselves. Local political renewal cannot be enough on its own. But it may be the important next step for individuals in breaking through the absurdity of the double life and for renewing democratic process. As people participate, they will begin to see why the facts of climate change matter to them and to develop a sociological imagination at the same time as they reconnect the rifts in time and space that have constructed climate change as a distant issue. As Eliasoph puts it, "With active, mindful political participation, we weave reality and a place for ourselves within it" (1998, 17). Working together may over time create the supportive community that is a necessary (though not sufficient) condition for people to face large fears about the future. Engagement in such activities may also serve an important strategy of providing hopeful action. According to Moser, working in a close community is important: "to help individuals stay engaged on an easily overwhelming issue, sort through complex issues, understand difficult trade-offs, and change habitual thoughts and behaviors, communicators must identify and engage sources of social support. Typically, interpersonal and small-group dialogue can address these needs much better than mass communication received in the privacy of one's living room" (Moser 2007a).

It will not be easy to overcome feelings of despair and ineffectiveness, to figure out how to communicate with neighbors across political differences, or to translate meaningfully the global into the local and vice versa. There is no guarantee that any of it will work. Facing climate change will not be easy. But it is worth trying.

Appendix A: Methods

Sundays are the traditional day for *Bygdabyingar* to head into the mountains for a *Sondagstur*, or Sunday hike. Without doubt, the most enjoyable aspect of my fieldwork in Bygdaby were the trips I took with the local Bygdaby hiking and skiing club. The *Bygdaby Utferdslag* was the largest of the local volunteer organizations in the community, and most of those who attended the hikes were members. But membership was not required; one simply showed up at the appointed time and place and paid a small fee for bus fare.

During the bus ride back from one of the more strenuous trips in which I had the pleasure of participating, I spoke with the woman in the seat beside me. She had just asked what I was studying. Before I could reply, a man sitting in front of me overheard our conversation and piped in with his own comment: "Don't study the fiddle!"—referring to the number of Americans and other foreigners who come to Norway with a fascination for traditional Norwegian culture. To which I replied, "I haven't even been there yet" (that is, to the local music academy in town).

"Good, you don't need to go."

"Don't you like fiddling?"

"It's not that; it's just that you should study something different."

Many foreign researchers have come to Norway drawn by fascination with Norwegian society—the high quality of life, environmental policies and values (Reed and Rothenberg 1993), and gender relations. Indeed, although I have been similarly impressed with and fascinated by these aspects of life in Norway for half my lifetime, my research questions are different from those that others pursue, as have been my methods. Rather than going to Norway to study my cultural roots in fiddling, to trace the ways environmental philosophy has developed from the Norwegian landscape, or to explore how patriarchal gender relations have been transformed under the welfare state, I have asked, *How are everyday people in a wealthy industrialized nation responding to the global environmental problem of climate change? Why are people not responding? What makes the issue difficult to think about? How do people manage to produce an everyday reality in which this critically serious problem is invisible? How do people in a wealthy, idealistic nation make sense of their position of global privilege?* And although I am fascinated with Norwegian tradition, I am also interested in how traditional activities ground the present day or form a sense of community and belonging, in how traditions are "invented" for present-day purposes (Hobsbawm and Ranger [1983] 1992), and how traditional practices may serve as "tools" that inform "strategies of action" (Swidler 1986). Two general questions guided my research on the public experience of climate change, culture, and everyday life in Bygdaby: What factors make it possible for people to think about large-scale environmental problems such as global warming? And what factors prevent or decrease this ability?

Why Norway?

Climate change poses serious ecological and political challenges to people worldwide. Why did I conduct my work in Norway, though? Can a study set in Norway add insight into a worldwide phenomenon? Especially because of the fact that this work offers a critique of a nation other than my own, it is important to emphasize that I did not choose to write about denial and privilege in Norway because I see Norwegians as somehow "worse" in this area or more morally at fault than Americans. Indeed, given the extreme levels of overconsumption and refusal to participate in the Kyoto process exhibited by my own native nation, such a critique would be absurd!

There is some element of chance that this research took place in Norway. My own personal history led me to live in Norway as an exchange student during high school. This experience fundamentally shaped my life, leading to a fascination with Norwegian society and people that has now spanned several decades. As a graduate student, I began to combine personal and professional interests in Norway and developed a different dissertation topic that would look into tradition and environmentalism in Norway. When my interests in global warming, denial, and privilege developed, I first thought I would switch to conducting research in the United States. However, I soon realized that Norway is a uniquely useful place to ask such questions. Norway shares features with other Western industrialized nations that exhibit denial of climate change (i.e., economic structure, political process, culture), yet it has an exceptional level of social justice and environmental consciousness, a highly informed and politically engaged population, and is situated in the Arctic, where the impacts of climate change are occurring at a faster rate.

Many of the issues that make it difficult to assess apathy in my own country, from economic to political, look very different in Norway. Norway has one of the highest levels of gross domestic product of any nation and a 50-year history of welfare state policies that has redistributed this wealth among the people. As a result, poverty on the scale found in the United States is virtually nonexistent. Finally, although there certainly existed skeptics about global warming at the time of my work in Bygdaby at the beginning of this century, such skepticism was much lower than in places such as the United States, where large counter-campaigns have been waged by industry-related organizations and the president himself openly questioned the validity of scientific data.

In terms of political activity, Norwegians again are exemplary. High percentages of Norwegians vote and are active in local politics. In this nation of 4.5 million inhabitants, the social distance between the "everyday person" and political leaders is much smaller than in the United States, providing a sense of accessibility to the political system. It is not uncommon for common Norwegian citizens to meet top political leaders such as the king or prime minister personally. And, in terms of generalizability, although Norway is also differentiated by variables of class, gender, and race, the nation's size, history, cultural traditions, and public policies have resulted in one of the most homogeneous nations in the world.[1] The fact that since the end of World War II Norwegians have been almost universally exposed to a defined, coherent set of information (through standardized school texts, government television and radio programs, and national media sources) makes it easier to distinguish denial from ignorance. Finally, Norwegians have been proud of their close relationships to nature, their native environmentalism, and their leadership role on global environmental issues, including climate change (Eriksen 1993, 1996). If any nation can find the ability to respond to this problem, it would be a place like Norway, where the population is educated, cared for, politicized, and environmentally engaged.

Examining nonresponse in Norway offers a kind of distilled version of dynamics that I suspect are present in other places, thus providing a way to isolate the phenomenon of collective avoidance by reducing complicating factors present in places such as the United States. Finally, as a researcher seeking to look at very basic features of society, I found it useful to work in a context other than in my own backyard, where so much can be taken for granted. My status as an outsider made it easier for me to ask basic questions about social life and practices and more acceptable for me to voice such questions to others.

The bulk of this project is based on one year of research in Norway, in 2000–2001, and includes 46 interviews, media analysis, and eight months of participant observation. In-depth exploration of relationships among thinking, feeling, performance, and meaning in everyday life can be accomplished only within a living context. I began with a prefieldwork phase in Oslo that lasted for several months and allowed me time to become immersed culturally, improve my language skills, and sensitize myself to important concepts through reading the newspaper, listening to the radio, talking with people, and generally moving through daily life. In Oslo, I also observed representations of culture and history (e.g., the Ski Museum, National Gallery Museum of Resistance during World War II, and others).

Fieldwork allows a researcher to gather general information about context and to approach an insider's view of reality. In my case, it allowed me to gather data in daily life about relationships among emotions, identity, meaning, and cultures of talk and about how people make sense of the issue of global warming. The coupling of participant observation with in-depth interviews allowed me access to different types of data by triangulating between methods, thereby increasing both validity and the kinds of data I could gather.

An Ethnography of the Invisible: The Community of Bygdaby

The people I spent time with live in a rural community in western Norway of about 14,000 inhabitants. Bygdaby is the commercial and educational center of the local region, with an economic base of commerce, tourism, industry, and agriculture. This town was a historically important farming community, and a relatively large proportion of the community still farms. Most farmers, however, also work second jobs in town—as teachers, bus drivers, merchants, and so forth. As is the case across most of Norway, Bygdaby has both modern features and strong roots in the past. The farm on which I lived had been in cultivation for roughly the past 4,000 years.

Within a week of my arrival in Bygdaby, I began attending local events in town: a march in memory of Crystal Night, the organizing meetings of the labor party, social events, a town meeting on a proposed hydro project, and so forth. The classifieds section of the local newspaper was an excellent source of information on local happenings. Other events were posted on billboards, or I learned of them through word of mouth. I began participant observation in the community in as wide a variety of public settings as time and access permitted (i.e., cafés, churches, schools, recreational organizations, work settings, town meetings, bars, and sporting events) in order to understand the range of variation within the community (i.e., maximum variation sampling). As I became familiar with the setting, and as general notions emerged, I concentrated on field sites that yielded the richest source of information on the themes I pursued: political events and activities, environmental issues, boundaries between insiders and outsiders, and the local sense of place and time.

During my stay in the community, I used my involvement in daily activities, careful observation, note taking, and, after a month or two, in-depth individual interviews to gain an understanding of how emotions, identity, meaning, and cultures of talk shaped how people in Bygdaby created a sense of everyday life as well as their perceptions and response to climate change. I observed public holidays and protest events and attended meetings of a range of active social and political organizations such as the Rotary Club, the Farm Women's Association, the Labor Party, and Amnesty International. In the last three months of my stay, I took a part-time position as a bartender at the historic hotel. Several evenings a week I worked either in the upstairs bar or the downstairs dance club making and serving drinks. This experience provided me with opportunities to interact informally with people in the community and to observe behavior in these settings. I made a number of interesting social observations while at work (i.e., about what people think and talk about while "relaxing" in these spaces), had useful conversations with community members, and made several contacts with people I later interviewed.

As a participant-observer, I attended to the kinds of things people talked about, how issues were framed, and, especially, topics that were not discussed. I watched regional television news and read the local and national newspapers, again paying attention to what was both present and missing and how the information presented was framed. In each setting, I paid particular attention to beliefs, emotions, and cultures of talk with respect to climate change: that is, whether climate change was discussed; if so, how it came up and how it was discussed; how people seemed to feel talking about it; and so forth (conversation and emotion norms, ways of framing information, how information was linked or not linked to other aspects of social life).[2] In this way, I began to understand in what spaces and ways it was acceptable to talk about climate change as well as what ideas were exchanged by different groups in different cultural spaces. I made nearly all my field notes in English. Although I did my best to remember the exact words of conversations in Norwegian, in most cases I was able to remember only the overall "gist." In such cases, I again used English to record what I remembered.

In addition to taking notes and observing, as a *participant*-observer I paid close attention to my own emotional responses and experiences as a source of information about the setting. My own feelings and reactions to the dynamics around me provided firsthand information about what it felt like to participate in different settings—although because I was both a foreigner and an outsider, my experiences were often different from those of local residents. As Norwegian-American researcher Anne Kiel notes, "It is one thing to speak the same language and an entirely different thing to speak the same culture" (1993a, 68).

An important part of understanding the organization of information at the local level was to watch how it entered the community in the form of news media (i.e., newspapers, radio, and TV). Throughout my time in Norway, I read the national Norwegian dailies (*Aftenposten, Dagbladet, VG*) as well as regional and local papers. This practice provided critical comparison not only for what information was available in each place, but for how issues were framed. I scanned papers focusing on climate change but remained open to other issues that also provided salient information. I filed newspaper articles of interest, highlighted portions of them, and made notes on their significance. In addition to collecting main articles on climate issues, I focused on letters to the editor and editorials. I periodically read U.S. newspapers through the Internet to obtain alternative perspectives on issues and to help me maintain an "outsider's mind" while living in Norway.

Site Selection and Representativeness

I felt that it was important to be in a smaller community in order to have access to a cross-section of the population and to be able to get a better overall sense of daily life. I selected Bygdaby for these reasons and because the residents spoke a dialect I was able to understand.³ Although there are relatively strong rural–urban and regional differences within Norway, I expect that readers familiar with Norway will recognize many of the features of life I describe to be similar to those in other parts of the country. To some extent, such commonalities change according to the type of material discussed. The social narratives I describe in chapter 5 are equally applicable nationwide. Sense of place, time, and community as well as traditional practices are more variable across Norway.

As a cultural outsider, I noticed larger features of society that most *Bygdabyingar* took for granted. The features I set in the foreground of my analysis exist in the background of people's lives. Some Norwegian readers may feel that by placing so much attention on minor background features of life, I have misunderstood what is really going on. However, these background features, unnoticed though they may be, organize daily life in deeply significant ways.

As an American doing research in Norway, I became aware of at least several underlying dynamics that influenced my interactions with the people I studied. First, there was the presence of *fremmed skeptis*, or the general tendency to be wary of people whom one does not know. Compared to Americans, Norwegians are more formal in social interactions and more likely to open up to people they meet through daily interaction or work settings than to "strangers" (see, e.g., Gullestad 1992). Even work colleagues can be guarded and maintain a protective bubble around fellow workers they don't know that well. One Saturday I met my job supervisor on the street with his wife and children. Edvard was my age and had always been friendly with me at work. Before and after work, four or so of us employees would sit and have coffee together. When I met him on the street, however, I received only a curt hello and nod as he walked by. He did not introduce me to his wife and kids. The next time I saw him at work he was normal and friendly. In the United States, such behavior would have been extremely rude. In Bygdaby, it was not at all uncommon. Although it is possible that Edvard simply did not like me and meant to snub me or even that he did like me and was therefore uncomfortable introducing me to his wife, I believe that the regularity of this type of "standoffish" behavior marks it as part of the larger norms of interaction that keenly divide insiders and outsiders in Norway. It is interesting to note that when I have discussed this issue with Norwegian friends (both those from Bygdaby and from elsewhere in Norway), at least three different people have explained to me that they are simply not used to meeting new people. Instead, relationships are usually based on longtime knowledge of one another. New relationships emerge slowly through work or school situations. And those whom one does not know personally are not acknowledged in chance interactions.

Features of life such as the fear of standing out and not speaking to strangers did not influence me so much as an *American* per se as they applied to me as a person who was unknown to the community members I spent time with. If anything, it seemed at times that my status as a foreigner exempted me from some of the suspicion I would otherwise have received, making it more normal for me to do and say some of the things that I did (because I was this type of outsider, I could hardly be expected to do things in the normal Norwegian or Bygdaby fashion). People explained things in more basic terms to me than I suspect they would to a native. As an outsider, I was a kind of cultural dope and thus less threatening.

Gaining Access

"And do they talk with you?" Margaret, an American woman who had married a Norwegian and had spent the past several years living in Norway with him and their kids, asked me about my research methods. Margaret and her husband, Paul, had come to visit Sam and me for the weekend for some skiing. We sat on my couch as their two-year-old explored the living room. Her question alluded to the degree of *fremmed skeptis*, guardedness toward unknown people, that is present in the Norwegian culture.

Despite my relative ability to gain access to the community, I was nonetheless clearly an outsider. This status had both advantages and disadvantages. On the one hand, a complete insider might not have been able to conduct the work I did. Being an American outsider made it more normal and acceptable to contact strangers, to ask questions about basic features of life in the town, and to break rules. On the other hand, as an outsider, however, I was never sure that I had completely understood what was happening, how gaps of both culture and language may have led me to misinterpretations and oversights. Meanings are lost in the translation of language. Gaps in understanding of emotions and of emotion norms may be particularly difficult to cross.

Past experiences of living in rural Norway (as an exchange student in a town of 4,000 in 1984–1985 and an employee in a fish factory in a town of 300 in 1987) provided me with some basis for understanding the circumstances I faced and made it easier for me to gain entry. I was familiar with the heightened degree of formality, the phenomenon of *fremmed skeptis*, codes of emotional expression and conduct in public space, and the way in which these norms are loosened when people drink. My entry into the community of Bygdaby, a rural Norwegian community, was made easier by my name, which is Norwegian, my status as a third-generation Norwegian American, and my level of fluency in the language.

The phenomenon of *fremmed skeptis* meant that the interviews I conducted with residents were sometimes on a more superficial level than I would have liked. Respondents were more careful and guarded than I have experienced when interviewing people in the United States. People were always pleasantly willing to talk with me, but I sometimes had the sense that they were presenting a particular face to an outsider.

I consciously took the role of the "novice," which in terms of Norwegian society and in many activities (back-country skiing, sausage making) I was. In addition, my language ability, which is strong but still limiting in group contexts, gave me a nonthreatening status. I believe that highlighting my status as an outside observer oftentimes gave me an honorary status and access to communities that would not otherwise have been open to me.

Interviews

In mid-January 2001, two months after beginning participant observation, I began conducting interviews. Anyone who has done interviews knows what an honor and privilege it can be to hear the stories of another person's life. The process of interviewing people in another culture adds to this level of fascination and wonder. It was a wonderful experience to listen to people tell about their lives, all the more so because the stories I was told were so different from my own or any story I had ever heard from a member of my own nation. I selected informants based on their (1) knowledge of local environmental politics, culture, or history; (2) sex, occupation, age, life experiences, or political background; (3) ability to articulate their views; and (4) personal rapport with me. I conducted in-depth, semistructured interviews with a total of 21 women and 25 men ranging in age from younger than 20 to older than 60 from a variety of occupations and from six of the eight active local political parties (see table A.1).

I contacted people for interviews whom I suspected had a variety of insights or experiences on my research questions. From each interview

Table A.1
Interview Respondent Characteristics

Characteristics	Number
Female	21
Male	25
Politically active	28
Not politically active	15
Unknown political proclivity	4
Younger than 20 years old	7
20–35 years old	8
35–60 years old	19
Older than 69	11
Labor Party member	8
Center Party member	2
Right Party member	1
Red Electoral Alliance (Communist) Party member	1
Socialist-Left Party member	3
Christian People's Party member	3
Unknown political affiliation	28
Student	7
Government employee	15
Business owner	5
Farmer (at least part-time)	4
Employee of nongovernmental organization	4
Employed in education	5
Other/unknown employment	6

came a range of answers, new ideas, and information. When interviewing people, I sometimes asked if they could suggest someone with very different views on this topic for me to interview in the future. I interviewed as wide a variety of people as I could find. They were farmers and students, businessmen and retired shopkeepers, as well as members of the various political parties. All but two interviews were conducted in Norwegian (the two exceptions were interviews with good English speakers). I tape recorded all interviews, transcribed them in Norwegian, and then translated key passages into English.

Interviews were focused on what it felt like to live in Bygdaby, how people created a sense of community, and the relationships between the community and the outside world. I asked people what they liked and did not like about their community, daily activities, their hopes and fears for their community, and their nation and the world in the future. I asked people what they felt were the most significant challenges faced by their community and their nation. I also asked them about activities in which I knew they were involved or concerned, and in this way I gathered information about other things that were happening in the community about which I otherwise would have been unaware.

If global warming was not raised as an issue naturally in the course of the interview (it often was not), I asked what people thought about the recent weather (which was widely described as abnormal) and followed with more specific questions on climate, whether they thought global warming was happening, if it troubled them, why or why not, whether or when they thought about it, whether they discussed it with family or friends, and so on. In all cases, I listened for how people thought about the wider world and how information about Norway's relationship to the wider world was filtered. As with participant observation, I watched for information on how people thought about global warming and other environmental problems, including the emotions, meanings, and identities they held and negotiated in doing such thinking. I also asked direct questions about the cultures of talk with respect to climate change and other environmental issues (although I relied on observation as my primary source of such information).

Providing an American Counterpoint

I supplemented the in-depth ethnographic account of what was happening in Bygdaby with a smaller sampling of voices from the United States. The United States is a large and politically, racially, and economically diverse nation. Conducting any study of denial and apathy in this country would be interesting but no doubt challenging to set up. Instead, what I present here is a sampling of ideas and observations that I believe expand the salience of the ethnographic material beyond the community in Norway. After my return to the United States, I was of course curious about similarities and differences. I made notes on discourses, events, and policies related to public perception of climate change. These experiences inform a background understanding of denial and political participation in an American context. Chapter 6 is also informed by secondary data on American experiences from psychologists and survey researchers. In 2008, I worked with a Whitman College student to conduct a set of ten interviews on climate change in a rural community near where I currently live in eastern Washington. The interviews followed very much the same format as those in Norway. Here, too, we were interested in how people make sense of climate change, what emotions are associated with the topic, and what norms and social spaces shape collective understanding of how and whether climate change is connected with everyday life. Throughout this time, my students submitted materials, made comments, and shared ideas with me about their own experiences. One day I more formally asked a group of students to describe their views. The experience, although again hardly representative of all Americans or even of all American students, was powerful for them and interesting for all of us, and it provided a provocative set of counterpoint voices in conversation with those from Bygdaby. All told, for the American context I offer ideas and voices but make no claim for representativeness. Readers will have to determine whether the shoe fits.

Appendix B: List of People in Bygdaby Interviewed and Quoted

Name	Age Category	Occupation/Other
Andre	early forties	unknown
Anne	midforties	farmer
Ann Marie	early fifties	teacher, lived in United States for some time
Arne	sixties	farmer, teacher
Åse	midforties	human rights activist
Audun	thirties	bartender
Bjarte	early sixties	farmer
Eirik	early fifties	county employee
Finn	fifties	farmer, unknown conversation at dance
Geir	forties	environmental activist from Oslo
Gunnar	early sixties	retired, led ski tours
Gunvar	thirties	bartender
Gurid	late twenties	student
Hans Olav	seventies	retired butcher
Harald	sixties	farmer
Hilde	sixties	farmer
Ingrid	about 15	high school student
Ingrid	late teens	student
Jan	fifties	unknown, conversation in bar
Joar	thirties	small business owner
Jon	late forties	unknown
Jorn	midforties	city administrator
Ketil	midfifties	administrator of a cultural institute
Kjell Magne	late forties	
Kjersti	early thirties	high school teacher
Knut	late fifties	county employee
Kristine	thirties	teacher
Lars	15	farmer's son
Lars	fifties	city councilor

(continued)

Name	Age Category	Occupation/Other
Lene	late forties	newspaper business
Lisbet	early forties	human rights activist
Lise	early forties	homemaker
Liv	thirties	human rights activist
Maghild	sixties	retired, unknown
Margit	fifties	
Marit	midthirties	
Martin	sixties	farmer
Mette	about 15	high school student
Mona	thirties	homemaker, city councilor
Øystein		
Per	fifties	taxi driver
Per Arne	early fifties	
Peter	forties	teacher
Sara	forties	
Sigurd	early thirties	shop keeper
Siri	about 15	high school student
Soren	late forties	county employee, environmental activist
Svein	early thirties	high school teacher
Torbjørn	early thirties	doctoral student
Torgeir	mid sixties	farmer
Torstein	midfifties	farmer
Trudi	about 15	high school student
Vigdis	17	student

Notes

Prologue

1. "Bygdaby" is a pseudonym. The word *bygdaby* in Norwegian, pronounced "Big-DAH-bee," is an expression that describes a community that is halfway between a rural hamlet (*bygd*) and city (*by*). I use the name *Bygdabyingar* to describe the residents of Bygdaby.

2. As noted in appendix A, I tape-recorded all interviews, transcribed them in Norwegian, and then translated key passages into English.

Introduction

1. This phenomenon is also reported in the United States (see, e.g., Immerwahr 1999).

2. In the climate negotiations, the Umbrella Group refers to an informal coalition of ten developed countries with strong economic and business interests in oil production that has most frequently lobbied to water down climate negotiations. The group evolved from the JUSSCANNZ group, which was active during the Kyoto Protocol.

Chapter 1

1. SOS Children's Village is a popular program in Norway that provides assistance to orphans in poor nations.

Chapter 2

1. On Crystal Night (November 9, 1938), more than 7,500 Jewish shops were destroyed, 400 synagogues were burned down, and 91 Jews were killed in Germany. An estimated 20,000 Jews were sent to concentration camps that had until this event been used mainly for political prisoners. The only people who were punished for crimes committed on Crystal Night were those who had raped

Jewish women, but only because they had broken the Nuremberg laws on sexual intercourse between Aryans and Jews.

2. This boundary around the local is what Opotow and Weiss call "temporal containment of harm." They classify it as a form of denial of outcome severity, "asserting that exposure to or injury from harms is an isolated, unlikely event rather than routine and/or chronic" (2000, 481).

3. In relation to this point, see the work of Cynthia Frantz and Stephan Mayer (2009), who apply a social psychological model of helping behavior to human response to climate change. Using this model, the authors identify five conditions that must be present in order for helping to occur. One of these conditions is that the potential helper must identify the event as an emergency situation. The authors note that the ambiguity of emergency situations may prevent people from responding and that people don't want to feel foolish by "overreacting." Individuals often navigate ambiguities by relying on the interpretations of others for their judgment of what is an emergency. The obvious application regarding denial of climate change is that because individuals often cannot see it directly, and because those around them are not reacting, they may not define it as an emergency.

Chapter 3

1. Norway has recently set an ambitious target to become carbon neutral by the year 2030. However, as many have pointed out, it is difficult to imagine how this goal can realistically be achieved under the current emissions scenario.

Chapter 4

1. I draw in particular on Anne Swidler's (1986) concept of the "cultural tool kit," Stanley Cohen's (2001) work on denial, Jocelyn Hollander and Hava Gordon's (2002) typology of thirteen "tools of social construction," Eviatar Zerubavel's (1997, 2002) work on "social dimensions of ignoring," and Nina Eliasoph's (1998) analysis of conversational tactics used to produce apathy.

Chapter 5

1. I have, for example, a Norwegian-language book for foreigners called *Typisk Norsk?* (Typical Norwegian?). The book cover features a light-skinned person with a Norwegian flag painted on his face—a common practice at international sporting events. Although the book describes a variety of tensions and complexities within Norway, it repeatedly presents the original stereotypes in both text and images (as in the act of using this image for the book cover).

2. There were more members in this age group probably also because younger people who spent time in the mountains were more likely to go on their own in less-organized groups.

3. Norway's footprint is actually the highest in Europe, but other nations are close: Sweden, 5.9 hectares per capita; France, 4.1; Germany, 5.3; Spain, 3.8; Philippines, 1.5; United States, 10.3. Data are from "Ranking the Ecological Impact of Nations," 1997.

Chapter 6

1. "Americans Consider Global Warming Real but not Alarming," is available at <http://www.gallup.com/poll/1822/americans-consider-global-warming-real-alarming.aspx> (accessed September 7, 2010).
2. "Global Warming on Public's Back Burner" is available at <http://www.gallup.com/poll/11398/Global-Warming-Publics-Back-Burner.aspx> (accessed September 7, 2010).

Chapter 7

1. In *Campos de Castilla* from 1912; the translation is from Machado 1978.

Appendix A

1. It is important to acknowledge that regional differences in character and perspective do exist, that rural–urban differences are strong, and that Norway is currently undergoing rapid cultural transformation, which is making for particularly sharp generational differences.
2. With this approach, I was able to study only *expressed* beliefs and emotions. That is, I was able to study beliefs and emotions only as they were presented to me in interviews and public space. If people did indeed have "real" emotions, these emotions may or may not have been the same as what they were willing to express publicly or to share with me or others in interviews and conversation.
3. Norway has two (closely related) languages and many dialects. The dialects are strongly varied enough that even native speakers from different parts of the country can have difficulty understanding one another.

References

Aall, Carlo, Kyrre Groven, and Gard Lindseth. 2007. The Scope of Action for Local Climate Policy: The Case of Norway. *Global Environmental Politics* 7 (2):83–101.

ACNielsen. 2007. Global Warming: A Self-Inflicted, Very Serious Problem, According to More Than Half the World's Online Population. <http://asiapacific.acnielsen.com/news/20070131.shtml> (accessed August 15, 2010).

Agarwal, A., and S. Narain. 1991. *Global Warming in an Unequal World: A Case of Environmental Colonialism*. New Delhi: Center for Science and Environment.

Agyeman, Julian, Peter Cole, Randolph Haluza-DeLay, and Pat O'Riley. 2009. *Speaking for Ourselves: Environmental Justice in Canada*. Toronto: University of British Columbia Press.

Alaska Regional Assessment Group. 1999. *The Potential Consequences of Climate Variability and Change Center for Global Change and Arctic System Research*. Fairbanks: University of Alaska. <http://www.besis.uaf.edu/regional-report/retional-report.html> (accessed August 13, 2010).

Anderson, Benedict. 1991. *Imagined Communities: Reflections on the Origin and Spread of Nationalism*. New York: Verso.

Arctic Climate Impact Assessment. 2005. *Arctic Climate Impact Assessment Scientific Report*. Cambridge: Cambridge University Press.

Arendt, Hannah. 1958. *The Human Condition*. Chicago: University of Chicago Press.

Arendt, Hannah. 1970. *On Revolution*. New York: Viking.

Arendt, Hannah. 1972. *Crises of the Republic*. New York: Harcourt Brace Jovanovich.

Armitage, Kevin. 2005. State of Denial. *United States and the Politics of Global Warming Globalizations* 2 (3):417–427.

Athanasiou, Tom, and Paul Baer. 2002. *Dead Heat: Global Justice and Climate Change*. New York: Seven Stories Press.

Baer, Paul, John Harte, Barbara Haya, Antonia Herzog, John Holdern, Nathan Hultman, Daniel Kammen, Richard Norgaard, and Leigh Raymond. 2000. Equity and Greenhouse Gas Responsibility. *Science* 289 (29):2287.

Bandura, Albert. 1997. *Self-Efficacy: The Exercise of Control*. New York: Freeman.

Barnett, T. P., J. C. Adam, and D. P. Lettenmaier. 2005. Potential Impacts of a Warming Climate on Water Availability in Snow-Dominated Regions. *Nature* 438:303–309.

Barstad, Anders., and Ottar Hellevik. 2004. På vei mot det gode samfunn? Om forholdet mellom ønsket og faktisk samfunnsutvikling (On Our Way to the Good Society? On the Relationship between Desired and Actual Social Evolution). Statistiske Analyser, 64. Available at <http://www.ssb.no/emner/00/02/sa64/> (accessed July 18, 2010).

Beck, Ulrich. 1992. *Risk Society: Towards a New Modernity*. Trans. Mark Ritter. London: Sage.

Beck, Ulrich. 2010. Remapping Social Inequalities in an Age of Climate Change: For a Cosmopolitan Renewal of Sociology. *Global Networks* 10 (2):165–181.

Bell, Alan. 1994. Climate of Opinion: Public and Media Discourse on the Global Environment. *Discourse & Society* 5 (1):33–64.

Bell, Michael. 1994. *Childerley: Nature and Morality in a Country Village*. Chicago: Chicago University Press.

Bellah, Robert, Richard Madsen, William M. Sullivan, Ann Swidler, and Steven M. Tipton. 2008. *Habits of the Heart: Individualism and Commitment in American Life*. Berkeley: University of California Press.

Bendelow, Gillian, and Simon Williams, eds. 1998. *Emotions in Social Life: Critical Themes and Contemporary Issues*. London: Routledge.

Berggreen, Brit. 1993. A National Identity in Person: The Making of a Modern Norwegian. In *Continuity and Change: Aspects of Contemporary Norway*, ed. Anne Cohel Kiel. Oslo: Scandinavia University Press, 39–54.

Borchgrevink, Tordis, and Øystein Gullvåg Holter, eds. 1995. *Labour of Love: Beyond the Self-Evidence of Everyday Life*. Aldershot: Avebury.

Bord, Richard, Ann Fisher, and Robert O'Connor. 1998. Public Perceptions of Global Warming: United States and International Perspectives. *Climate Research* 11:75–84.

Bord, Richard, Robert E. O'Connor, and Ann Fischer. 2000. In What Sense Does the Public Need to Understand Global Climate Change? *Public Understanding of Science (Bristol, England)* 9 (3):205–218.

Bostrom, Ann, M. Granger Morgan, Baruch Fischoff, and Daniel Read. 1994. What Do People Know about Global Climate Change? I. Mental Models. *Risk Analysis* 14 (6):959–970.

Boykoff, Maxwell. 2008a. The Cultural Politics of Climate Change: Discourse in UK Tabloids. *Political Geography* 27:549–569.

Boykoff, Maxwell. 2008b. Lost in Translation? The United States Television News Coverage of Anthropogenic Climate Change, 1995–2004. *Climatic Change* 86:1–11.

Boykoff, Maxwell T., and Jules M. Boykoff. 2004. Balance as Bias: Global Warming and the US Prestige Press. *Global Environmental Change Part A* 14:125–136.

Brechin, Steven. 2003. Comparative Public Opinion and Knowledge on Global Climatic Change and the Kyoto Protocol: The U.S. versus the World? *International Journal of Sociology and Social Policy* 23 (10):106–134.

Brechin, Steven R. 2008. Ostriches and Change: A Response to "Global Warming and Sociology." *Current Sociology* 56 (3):467–474.

Brewer, Thomas. 2005. U.S. Public Opinion on Climate Change Issues: Implications for Consensus Building and Policymaking. *Climate Policy* 4:359–376.

Brick, Philip, Donald Snow, and Sarah van de Wetering. 2001. *Across the Great Divide: Explorations in Collaborative Conservation and the American West.* Washington, DC: Island Press.

Brossard, Dominique, James Shanahan, and Katherine McComas. 2004. Are Issue-Cycles Culturally Constructed? A Comparison of French and American Coverage of Global Climate Change. *Mass Communication & Society* 7 (3):359–377.

Brulle, Robert J. 2010. From Environmental Campaigns to Advancing the Public Dialog: Environmental Communication for Civic Engagement. *Environmental Communication: A Journal of Nature and Culture* 4 (1):82–98.

Bryn, Steinar. 1994. The Americanization of Norwegian Culture. PhD diss., Department of Philosophy, University of Minnesota.

Buell, Frederick. 2003. *From Apocalypse to Way of Life: Environmental Crisis in the American Century.* New York: Routledge.

Bulkeley, Harriet. 2000. Common Knowledge? Public Understanding of Climate Change in Newcastle, Australia. *Public Understanding of Science* 9 (3):313–333.

Campbell-Lendrum, D., A. Pruss-Ustun, and C. Corvalan. 2003. How Much Disease Could Climate Change Cause? In *Climate Change and Health: Risks and Responses*, ed. D. Campbell-Lendrum, C. Corvalan, K. L. Ebi, A. K. Githeko, and J. S. Scheraga. Geneva, Switzerland: World Health Organization.

Carlson, Ann E. 2007. Heat Waves, Global Warming, and Mitigation. *Issues in Legal Scholarship, Catastrophic Risks: Prevention, Compensation, and Recovery*. <http://www.bepress.com/ils/iss10/art7> (accessed July 18, 2010).

Carvalho, Anabela. 2005. Representing the Politics of the Greenhouse Effect: Discursive Strategies in the British Media. *Critical Discourse Studies* 2 (1):1–29.

Chomsky, Noam. 2002. *Media Control: The Spectacular Achievements of Propaganda.* New York: Seven Stories Press.

Cohen, Stanley. 2001. *States of Denial: Knowing about Atrocities and Suffering*. Cambridge, MA: Polity Press.

Crossley, Nick. 1997. Emotion and Communicative Action: Habermas, Linguistic Philosophy, and Existentialism. In *Emotions in Social Life: Critical Themes and Contemporary Issues*, ed. Gillian Bendelow and Simon J. Williams. New York: Routledge, 16–38.

Denzin, Norman. 1984. *On Understanding Emotions*. San Francisco: Jossey-Bass.

Dilling, Lisa, and Susanne Moser. Forthcoming. Communicating Science to the Public. In *Oxford Handbook of Climate Change and Society*, ed. John Dryzek, David Schlosberg, and Richard Norgaard. Oxford, UK: Oxford University Press.

Dispensa, Jaclyn Marissa, and Robert J. Brulle. 2003. Media's Social Construction of Environmental Issues: Focus on Global Warming—a Comparative Study. *International Journal of Sociology and Social Policy* 23 (10):74–105.

Donohoe, M. 2003. Causes and Health Consequences of Environmental Degradation and Social Injustice. *Social Science & Medicine* 56 (3):573–587.

Dunlap, Riley. 1998. Lay Perceptions of Global Risk: Public Views of Global Warming in Cross National Context. *International Sociology* 13 (4):473–498.

Dunwoody, Sharon. 2007. The Challenge of Trying to Make a Difference Using Media Messages. In *Creating a Climate for Change: Communicating Climate Change and Facilitating Social Change*, ed. Susanne C. Moser and Lisa Dilling. New York: Cambridge University Press.

Durkheim, Emile. 1915. *The Elementary Forms of the Religious Life*. London: Allen and Unwin.

Eliasoph, Nina. 1998. *Avoiding Politics: How Americans Produce Apathy in Everyday Life*. Cambridge, UK: Cambridge University Press.

Eriksen, Thomas Hylland. 1993. Being Norwegian in a Shrinking World. In *Continuity and Change: Aspects of Contemporary Norway*, ed. Anne Cohel Kiel. Oslo: Scandinavia University Press, 11–38.

Eriksen, Thomas Hylland. 1996. Norwegians and Nature. From official Norwegian government Web site at <http://folk.uio.no/geirthe/Nature.html> (acessed July 18, 2010).

Festinger, Leon. 1957. *A Theory of Cognitive Dissonance*. Evanston, IL: Row, Peterson.

Frantz, Cynthia, and Stephan Mayer. 2009. The Emergency of Climate Change: Why Are We Failing to Take Action? *Analyses of Social Issues and Public Policy* 9 (1): 205–222.

Frederick, K. D., and D. C. Major. 1997. Climate Change and Water Resources. *Climate Change* 37:7–23.

Freudenberg, William. 2005. Privileged Access, Privileged Accounts: Toward a Socially Structured Theory of Resources and Discourses. *Social Forces* 94 (1):89–114.

Freudenburg, William, Robert Gramling, and Debra J. Davidson. 2008. Scientific Certainty Argumentation Methods (SCAMs): Science and the Politics of Doubt. *Sociological Inquiry* 78 (1):2–38.

Gallup News Service. 2010. Trust in Government. <http://www.gallup.com/poll/5392/Trust-Government.aspx> (accessed August 16, 2010).

Gamson, William. 1992. *Talking Politics*. New York: Cambridge University Press.

Gecas, Victor, and Peter Burke. 1995. Self and Identity. In *Sociological Perspective on Social Psychology*, ed. Karen Cook, Gary Alan Fine, and James House. Boston: Allyn and Bacon, 41–67.

Giddens, Anthony. 1991. *Modernity and Self Identity: Self and Society in the Late Modern Age*. Cambridge, UK: Polity Press.

Giddens, Anthony. 2009. *The Politics of Climate Change*. London: Polity Press.

GlobeScan. 1999. Environics International Environmental Monitor Survey Dataset. <http://jeff-lab.queensu.ca/poadata/analyze/iem/iem.shtml> (accessed April 14, 2007).

Goffman, Erving. 1959. *The Presentation of Self in Everyday Life*. New York: Anchor.

Goffman, Erving. 1974. *Frame Analysis: An Essay on the Organization of Experience*. New York: Harper and Row.

Goodwin, Jeff, James M. Jasper, and Francesca Polletta. 2001. *Passionate Politics: Emotions in Social Movements*. Chicago: University of Chicago Press.

Gramsci, Antonio. 1971. *The Prison Notebooks*. New York: International Publishers.

Grundmann, Reiner. 2006. Ozone and Climate: Scientific Consensus and Leadership. *Science, Technology, & Human Values* 31 (1):73–101.

Grundmann, Reiner. 2007. Climate Change and Knowledge Politics. *Environmental Politics* 16 (3):414–432.

Gullestad, Marianne. 1992. *The Art of Social Relations: Essays on Culture, Social Action, and Everyday Life in Modern Norway*. Oslo: Scandinavian University Press.

Gullestad, Marianne. 1997. A Passion for Boundaries: Reflections on the Connection between the Everyday Lives of Children and Discourses on the Nation in Norway. *Childhood* 4 (1):19–42.

Habermas, Jürgen. 1984. *Reason and the Rationalization of Society*. Vol. I. *The Theory of Communicative Action*. Boston: Beacon Press.

Hagen, Kåre, and Jon Hippe. 1993. The Norwegian Welfare State: From Postwar Consensus to Future Conflicts? In *Continuity and Change: Aspects of Contemporary Norway*, ed. Anne Cohel Kiel. Oslo: Scandinavia University Press, 85–105.

Hales, Simon, N. de Wet, J. Maindonald, and A. Woodward. 2002. Potential Effect of Population and Climate Changes on Global Distribution of Dengue Fever: An Empirical Model. *Lancet* 360: 830–834.

Halford, Grame, and Peter Sheehan. 1991. Human Responses to Environmental Changes. *International Journal of Psychology* 26 (5):599–611.

Hamilton, L. C., D. E. Rohall, C. Brown, G. Hayward, and B. D. Keim. 2003. Warming Winters and New Hampshire's Lost Ski Areas: An Integrated Case Study. *International Journal of Sociology and Social Policy* 23:52–73.

Harrington, Julie, and Todd Walton. 2007. Climate Change in Coastal Areas in Florida: Sea Level Rise Estimation and Economic Analysis to Year 2080. Center for Economic Forecasting and Analysis, Florida State University. <http://www.cefa.fsu.edu/uploaded%20current%20projects/ncep.pdf> (accessed August 16, 2010).

Haugestad, Anne Kristine. 2003. Norwegians as Global Neighbors: A Presentation of the Fair Share Approach to Globally Responsible Consumption. Paper presented to the Second Global Conference: Ecological Justice and Global Citizenship, February 2003, Copenhagen.

Hellevik, Ottar. 2002. Beliefs, Attitudes, and Behavior towards the Environment. In *Realizing Rio in Norway: Evaluative Studies of Sustainable Development*, ed. William Lafferty, Morton Nordskog, and Hilde Annette Aakre. Oslo: Program for Research and Documentation for a Sustainable Society (Prosus), University of Oslo, 7–19.

Hellevik, Ottar, and Henning Høie. 1999. Vi bekymrer oss mindre for miljøet. *Samfunnsspeilet* 13 (4):53–61.

Herman, Edward, and Noam Chomsky. 2002. *Manufacturing Consent: The Political Economy of the Mass Media*. New York: Pantheon Books.

Hille, Jon. 1995. *Sustainable Norway*. Oslo: Norwegian Forum for Development and Environment.

Hobsbawm, Eric, and Terence Ranger. [1983] 1992. *The Invention of Tradition*. New York: Cambridge University Press.

Hochschild, Arlie Russell. 1983. *The Managed Heart: Commercialization of Human Feeling*. Berkeley: University of California Press.

Hollander, Jocelyn A., and Hava R. Gordon. 2006. The Processes of Social Construction in Talk. *Symbolic Interaction* 29 (2):183–212.

Hovden, Eivind, and Gard Lindseth. 2002. Norwegian Climate Policy 1989–2002. In *Realizing Rio in Norway: Evaluative Studies of Sustainable Development*, ed. William Lafferty, Morton Nordskog, and Hilde Annette Aakre. Oslo: Program for Research and Documentation for a Sustainable Society (Prosus), University of Oslo, 143–168.

Hovden, Eivind, and Gard Lindseth. 2004. Discourses in Norwegian Climate Policy: National Action or Thinking Globally? *Political Studies* 52 (1):63–81.

Immerwahr, John. 1999. Waiting for a Signal: Public Attitudes toward Global Warming, the Environment, and Geophysical Research. Washington, DC: Public Agenda/American Geophysical Union. <http://www.agu.org> (no longer available).

Intergovernmental Panel on Climate Change (IPCC). 1990. *First Assessment Report*. Cambridge: Cambridge University Press for the IPCC.

Intergovernmental Panel on Climate Change (IPCC). 1995. *Second Assessment Report*. Cambridge: Cambridge University Press for the IPCC.

Intergovernmental Panel on Climate Change (IPCC). 2001. *Third Assessment Report*. Cambridge: Cambridge University Press for the IPCC.

Jacoby, Susan. 2008. *The Age of American Unreason*. New York: Pantheon.

Jacques, Peter. 2006. The Rearguard of Modernity: Environmental Skepticism as a Struggle of Citizenship. *Global Environmental Politics* 6:1.

Jacques, Peter. 2009. *Environmental Skepticism: Ecology, Power, and Public Life*. Burlington, VT: Ashgate.

Jacques, Peter, Riley Dunlap, and Mark Freeman. 2008. The Organisation of Denial: Conservative Think Tanks and Environmental Skepticism. *Environmental Politics* 17 (3):349–385.

Jaeger, Carlo, Ortwin Renn, Eugene Rosa, and Thomas Webler. 2001. *Risk, Uncertainty, and Rational Action*. London: Earthscan.

Jasper, James. 1997. *The Art of Moral Protest*. Chicago: University of Chicago Press.

Jasper, James M. 1998. The Emotions of Protest: Affective and Reactive Emotions in and around Social Movements. *Sociological Forum* 13 (3):397–424.

Jonassen, Christen. 1983. *Value Systems and Personality in a Western Civilization: Norwegians in Europe and America*. Columbus: Ohio State University Press.

Katzenstein, Peter. 1985. *Small States in World Markets: Industrial Policy in Europe*. Ithaca, NY: Cornell University Press.

Kellstedt, Paul, Sammy Zahran, and Arnold Vedlitz. 2008. Personal Efficacy, the Information Environment, and Attitudes toward Global Warming and Climate Change in the United States. *Risk Analysis* 28 (1):113–126.

Kempton, Willett, James S. Boster, and Jennifer A. Hartley. 1996. *Environmental Values in American Culture*. Cambridge, MA: MIT Press.

Kent, Jennifer. 2009. Individualized Responsibility and Climate Change: "If Climate Protection Becomes Everyone's Responsibility, Does It End Up Being No-one's?" *Cosmopolitan Civil Societies Journal* 3 (1):132–149.

Kiel, Anne Cohel. 1993a. Confessions of an Angry Commuter: Or Learning How to Communicate the Non-Communicating Way. In *Continuity and Change: Aspects of Contemporary Norway*, ed. Anne Cohel Kiel, 55–70. Oslo: Scandinavia University Press.

Kiel, Anne Cohel, ed. 1993b. *Continuity and Change: Aspects of Contemporary Norway*. Oslo: Scandinavia University Press.

Kimmel, Michael S., and Abby L. Ferber. 2003. *Privilege: A Reader*. Boulder, CO: Westview Press.

Klandermas, Bert. 1997. *The Social Psychology of Protest*. Cambridge, UK: Blackwell.

Klausen, Arne Martin. 1984. *Den Norske væremåten*. Oslo: Cappelen.

Klima og Forurensnings Direktoratat. 2009. *Utslipp fra olje- og gassindustrien 2009*. <http://www.klif.no/no/Aktuelt/Nyheter/2010/Juni-2010/Utslipp-fra-olje-og-gassindustrien-2009> (accessed August 12, 2010).

Kollmuss, Anya, and Julian Agyeman. 2002. Mind the Gap: Why Do People Act Environmentally and What Are the Barriers to Pro-environmental Behavior? *Environmental Education Research* 8 (3):239–260.

Krosnick, Jon. 2009. The Associated Press–Stanford University Environment Poll. <http://www.apgfkpoll.com/pdf/APStanford_University_Environment_Poll_Topline.pdf> (accessed July 18, 2010).

Krosnick, Jon, Allyson Holbrook, Laura Lowe, and Penny Visser. 2006. The Origins and Consequences of Democratic Citizen's Policy Agendas: A Study of Popular Concern about Global Warming. *Climatic Change* 77: 7–43.

Lafferty, William, Jørgen Knudsen, and Olav Mosvold Larsen. 2007. Pursuing Sustainable Development in Norway: The Challenge of Living up to Brundtland at Home. *European Environment* 17(3): 177–188.

Lafferty, William M., and James R. Meadowcroft, eds. 2000. *Implementing Sustainable Development Strategies and Initiatives in High Consumption Societies*. Oxford, UK: Oxford University Press.

Langhelle, Oluf S. 2000. Norway: Reluctantly Carrying the Torch. In *Implementing Sustainable Development: Strategies and Initiatives in High Consumption Societies*, ed. W. M. Lafferty and J. Meadowcroft. Oxford: Oxford University Press, 174–208.

Larsen, P., O. S. Goldsmith, O. Smith, M. Wilson, K. Strzepek, P. Chinowsky, and B. Saylor. 2008. *Estimating the Future Costs of Alaska Public Infrastructure at Risk to Climate Change*. East Anglia: Elsevier Press.

Lean, Geoffrey. 2005. Apocalypse Now: How Mankind Is Sleepwalking to the End of the Earth: Floods, Storms, and Droughts. *The Independent* (February). <http://news.independent.co.uk/world/environment/article13813.ece> (accessed July 19, 2010).

Leiserowitz, Anthony. 2005. American Risk Perceptions: Is Climate Change Dangerous? *Risk Analysis* 25 (6):1433–1442.

Leiserowitz, Anthony. 2006. Climate Change Risk Perception and Policy Preferences: The Role of Affect, Imagery, and Values. *Climatic Change* 77 (1):45–72.

Leiserowitz, Anthony. 2007. Communicating the Risks of Global Warming: American Risk Perceptions, Affective Images, and Interpretive Communities. In *Creating a Climate for Change: Communicating Climate Change and Facilitating Social Change*, ed. Susanne C. Moser and Lisa Dilling. New York: Cambridge University Press, 44–63.

Leiserowitz, Anthony, Edward Maibach, and Connie Roser-Renouf. 2008. *Global Warming's "Six Americas": An Audience Segmentation.* New Haven, CT, and Fairfax, VA: Yale Project on Climate Change, School of Forestry, and Environmental Studies, Yale University, and Center for Climate Change Communication, George Mason University.

Leiserowitz, Anthony, Edward Maibach, and Connie Roser-Renouf. 2010. *Climate Change in the American Mind: Americans' Global Warming Beliefs and Attitudes in January 2010.* New Haven, CT, and Fairfax, VA: Yale Project on Climate Change, Yale University, and Center for Climate Change Communication, George Mason University. <http://environment.yale.edu/uploads/Americans GlobalWarmingBeliefs2010.pdf> (accessed August 13, 2010).

Lertzman, Renee. 2008. The Myth of Apathy. *Ecologist* 34 (4).

Lertzman, Renee. 2009. The Myth of Apathy: Psychosocial Dimensions of Environmental Degradation. PhD diss., Doctor of Philosophy of Cardiff University, School of Social Sciences, December.

Liévanos, Raoul S. 2010. A Minority Perspective Is Limited: Environmental Privilege and Surface Water Hazards in an Impaired Estuary. Presented at the annual meeting of the Association of American Geographers, April 18, Washington, DC.

Lifton, Robert J. 1995. *Hiroshima in America: Fifty Years of Denial.* New York: G. P. Putnam Sons.

Lifton, Robert J. 1967. *Death in Life: Survivors of Hiroshima.* New York: Random House.

Lifton, Robert J. 1982. *Indefensible Weapons: The Political and Psychological Case against Nuclearism.* New York: Basic Books.

Lifton, Robert J. 1993. *The Protean Self: Human Resiliency in an Age of Fragmentation.* New York: Basic Books.

Lindseth, Gard. 2006. Scalar Strategies in Climate-Change Politics: Debating the Environmental Consequences of a Natural Gas Project. *Environment and Planning. C, Government & Policy* 24 (5):739–754.

Lorenzoni, Irene, Anthony Leiserowitz, Doria Migue De Franca, Wouter Poortinga, and Nick Pidgeon. 2006. Cross-National Comparisons of Image Associations with "Global Warming" and "Climate Change" among Laypeople in the United States of America and Great Britain. *Journal of Risk Research* 9 (3):265–281.

Lorenzoni, Irene, Sophie Nicholson-Cole, and Lorraine Whitmarsh. 2007. Barriers Perceived to Engaging with Climate Change among the UK Public and Their Policy Implications. *Global Environmental Change* 17 (3–4):445–459.

Lorenzoni, Irene, and Nick F. Pidgeon. 2006. Public Views on Climate Change: European and USA Perspectives. *Climatic Change* 77:73–95.

Lukes, Steven. 1974. *Power: A Radical View.* New York: Macmillan.

Machado, Antonio. 1978. *Selected Poems of Antonio Machado.* Baton Rouge: Louisiana State University Press.

Macnaghten, P. 2003. Embodying the Environment in Everyday Life Practices. *Sociological Review* 51 (1):63–84.

Macy, Joanna. 1991. *World as Lover, World as Self.* Berkeley, CA: Parallax.

Macy, Joanna, and Molly Young Brown. 1998. *Coming Back to Life.* Gabriola Island, Canada: New Society.

Maniates, Michael. 2002. Individualization: Plant a Tree, Buy a Bike, Save the World? In *Confronting Consumption*, ed. T. Princen, M. Maniates, and K. Conca. Cambridge, MA: MIT Press.

Maslow, Abraham H. 1970. *Motivation and Personality.* New York: Harper & Row.

McCarthy, Michael. 2007. We Shouldn't Be Shocked by Public Apathy. The Independent, July 6. <http://www.independent.co.uk/environment/climate-change/we-shouldnt-be-shocked-by-public-apathy-456155.html> (accessed July 25, 2010).

McCright, Aaron M., and Riley E. Dunlap. 2003. Defeating Kyoto: The Conservative Movement's Impact on U.S. Climate Change Policy. *Social Problems* 50:348–373.

McCright, Aaron M., and Riley E. Dunlap. 2000. Challenging Global Warming as a Social Problem: An Analysis of the Conservative Movement's Counter-claims. *Social Problems* 47:499–522.

Meijnders, Anneloes L., Cees J. H. Midden, and Henk A. M. Wilke. 2001a. Communications about Environmental Risks and Risk-Reducing Behavior: The Impact of Fear on Information Processing. *Journal of Applied Social Psychology* 31 (4):754–777.

Meijnders, Anneloes L., Cees J. H. Midden, and Henk A. M. Wilke. 2001b. Role of Negative Emotions in Communication about $CO2$ Risks. *Risk Analysis* 21 (5):955–966.

Michaels, David, and Celeste Monforton. 2005. Manufacturing Uncertainty: Contested Science and the Protection of the Public's Health and Environment. *American Journal of Public Health* 95 (S1):S39–S48.

Miller, Arthur H., and Ola Listhaug. 1998. Policy Preferences and Political Distrust: A Comparison of Norway, Sweden, and the United States. *Scandinavian Political Studies* 21 (2):161–187.

Mills, C. W. 1959. *The Sociological Imagination.* New York: Oxford University Press.

Mjøset, Lars. 1993. Norwegian Political Economy. In *Continuity and Change: Aspects of Contemporary Norway*, ed. Anne Cohel Kiel. Oslo: Scandinavia University Press, 107–131.

Mooney, Christopher. 2005. *The Republican War on Science.* New York: Basic Books.

Moser, Susanne. 2007a. In the Long Shadows of Inaction: The Quiet Building of a Climate Protection Movement in the United States. *Global Environmental Politics* 7 (2):124–144.

Moser, Susanne. 2007b. More Bad News: The Risk of Neglecting Emotional Responses to Climate Change Information. In *Creating a Climate for Change: Communicating Climate Change and Facilitating Social Change*, ed. Susanne C. Moser and Lisa Dilling. New York: Cambridge University Press, 64–80.

Moser, Susanne C., and Lisa Dilling. 2004. Making Climate Hot: Communicating the Urgency and Challenge of Global Climate Change. *Environment* 46 (10):32–46.

Moser, Susanne C., and Lisa Dilling, eds. 2007. *Creating a Climate for Change: Communicating Climate Change and Facilitating Social Change*. Cambridge, UK: Cambridge University Press.

Moser, Susanne, and Maggie Walser. 2008. Communicating Climate Change Motivating Citizen Action. In *Encyclopedia of Earth*, ed. Cutler J. Cleveland. Washington, DC: Environmental Information Coalition, National Council for Science and the Environment. <http://www.eoearth.org/article/Communicating_climate_change_motivating_citizen_action> (accessed August 16, 2010).

Newport, Frank. 2008. Little Increase in Americans' Global Warming Worries. Princeton, NJ: Gallup. <http://www.gallup.com/poll/106660/little-increase-americans-global-warming-worries.aspx> (accessed August 10, 2010).

Nicholsen, Shierry. 2002. *The Love of Nature and the End of the World*. Cambridge, MA: MIT Press.

Niemi, Richard G., and Herbert F. Weisberg. 2001. *Controversies in Voting Behavior*. 4th ed. Washington, DC: CQ Press.

Nilsen, Ann. 1999. Where Is the Future? Time and Space as Categories in Analyses of Young People's Images of the Future. *Innovation* 12 (2):175–194.

Nisbet, Matthew, and Teresa Myers. 2007. The Polls—Trends: Twenty Years of Public Opinion about Global Warming. *Public Opinion Quarterly* 71 (3):444–470.

Norgaard, Kari Marie. 1999. Moon Phases, Menstrual Cycles, and Mother Earth: The Construction of a Special Relationship between Women and Nature. *Ethics and the Environment* 4 (2):197–209.

Norgaard, Kari Marie. 2006a. "People Want to Protect Themselves a Little Bit": Emotions, Denial, and Social Movement Nonparticipation. *Sociological Inquiry* 76:372–396.

Norgaard, Kari. 2006b. We Don't Really Want to Know": The Social Experience of Global Warming: Dimensions of Denial and Environmental Justice. *Organization & Environment* 19 (3):347–470.

Norgaard, Kari Marie. 2009. *Cognitive and Behavioral Challenges in Responding to Climate Change*. Policy Research Working Paper no. 4940. Washington, DC: World Bank.

Norgaard, Kari Marie. Forthcoming. Climate Denial: Emotion, Psychology, Culture and Political Economy. In *Oxford Handbook of Climate Change and Society*, ed. John Dryzek, David Schlosberg, and Richard Norgaard. Oxford, UK: Oxford University Press.

Norgaard, Kari, and Alan Rudy. 2008. Climate Change and the Sociological Imagination. *Footnotes: Newsletter of the American Sociological Association.* <http://www.asanet.org/footnotes/dec08/climate.html>.

Norwegian Ministry of Petroleum and Energy (MoPE). 2002. Fact Sheet 2002 Norwegian Petroleum Activity. <http://www.regjeringen.no/en/dep/oed/dok/veiledninger_brosjyrer/2002/Fact-Sheet-2002-Norwegian-Petroleum-Activity-.html?id=419395> (accessed August 13, 2010).

Nutter, Franklin W., and Reinsurance Association of America. 2007. Testimony before the U.S. House Select Committee on Energy Independence and Global Warming on the Economic Impacts of Global Warming. 110th Cong., 1st sess., May 3, 2007.

O'Connor, Robert, Richard J. Bord, Brent Yarnal, and Nancy Wiefek. 2002. Who Wants to Reduce Greenhouse Gas Emissions? *Social Science Quarterly* 83 (1):1–17.

Opotow, Susan, and Leah Weiss. 2000. New Ways of Thinking about Environmentalism: Denial and the Process of Moral Exclusion in Environmental Conflict. *Journal of Social Issues* 56 (3):475–490.

Pettit, Jethro. 2004. Climate Justice: A New Social Movement for Atmospheric Rights. *IDS Bulletin* 35 (3):102–106.

Pew Research Center for the People and the Press. 2009. *Economy, Jobs Trump All Other Policy Priorities in 2009: Environment, Immigration, Health Care Slip Down the List.* Washington, DC: Pew Research Center for the People and the Press.

Polletta, Francesca. 1998. It Was Like a Fever: Narrative and Identity in Collective Action. *Social Problems* 45:137–159.

Pulido, Laura. 2000. Rethinking Environmental Racism: White Privilege and Urban Development in Southern California. *Annals of the Association of American Geographers. Association of American Geographers* 90: 12–40.

Putnam, Robert D. 2000. *Bowling Alone: The Collapse and Revival of American Community.* New York: Simon & Schuster.

Ranking the Ecological Impact of Nations. 1997. <http://www.ecouncil.ac.cr/rio/focus/report/english/footprint/ranking.htm> (accessed December 2002).

Räthzel, N., and D. Uzzell. 2009. Changing Relations in Global Environmental Change. *Global Environmental Change* 19 (3):326–335.

Read, Daniel, Ann Bostrom, M. Granger Morgan, Baruch Fischoff, and Tom Smuts. 1994. What Do People Know about Global Climate Change? II. Survey Studies of Educated Lay People. *Risk Analysis* 14 (6):971–982.

Reed, Peter, and David Rothenberg. 1993. *Wisdom in the Open Air: The Norwegian Roots of Deep Ecology.* Minneapolis: University of Minnesota Press.

Roberts, J. Timmons. 2001. Global Inequality and Climate Change. *Society & Natural Resources* 14 (6):501–509.

Roberts, J. Timmons, and Bradley C. Parks. 2007. *A Climate of Injustice: Global Inequality, North–South Politics, and Climate Policy*. Cambridge, MA: MIT Press.

Robine, J. M., S. L. Cheung, S. Le Roy, H. Van Oyen, C. Griffiths, J. P. Michel, and F. R. Herrmann. 2008. Death Toll Exceeded 70,000 in Europe during the Summer of 2003. *Comptes Rendus Biologies* 331:171–178.

Roosevelt, Margo. 2004. Vanishing Alaska. *Time Magazine*, October 4. <http://www.time.com/time/magazine/article/0,9171,995264,00.html> (accessed August 13, 2010).

Rosa, Eugene. 2001. Global Climate Change: Background and Sociological Contributions. *Society & Natural Resources* 6 (14):491–499.

Rosenberg, Morris. 1991. Self-Processes and Emotional Experiences. In *The Self-Society Dynamic: Cognition, Emotion, and Action*, ed. Judith Howard and Peter Callero. Cambridge, UK: Cambridge University Press, 123–142.

Roy, Arundhati. 2005. Public Power in the Age of Empire. Keynote address, American Sociological Association, San Francisco, CA, August 16.

Rund, Audun. 2002. Industry and Environmental Responsibility: From Proactive to Reactive Public Policies. In *Realizing Rio in Norway: Evaluative Studies of Sustainable Development*, ed. William Lafferty, Morton Nordskog, and Hilde Annette Aakre. Oslo: Program for Research and Documentation for a Sustainable Society (Prosus), University of Oslo, 63–86.

Saad, Lydia. 2002. Americans Sharply Divided on Seriousness of Global Warming. Gallup News Service, March. <http://www.gallup.com/poll/5509/Americans-Sharply-Divided-Seriousness-Global-Warming.aspx> (accessed August 13, 2010).

Saad, Lydia. 2007. Did Hollywood's Glare Heat Up Public Concern about Global Warming? Concern about Global Warming Is Up Slightly over Past Year. Gallup News Service, March 21. <http://www.gallup.com/poll/26932/did-hollywoods-glare-heat-public-concern-about-global-warming.aspx> (accessed August 13, 2010).

Sandvik, Hanno. 2008. Public Concern over Global Warming Correlates Negatively with National Health. *Climatic Change* 90 (3):333–341.

Scheff, Thomas. 1988. Shame and Conformity: The Deference–Emotion System. *American Sociological Review* 53:395–406.

Scheff, Thomas. 1990. *Microsociology*. Chicago: University of Chicago Press.

Scheff, Thomas. 1997. *Emotions, the Social Bond, and Human Reality: Part/Whole Analysis*. New York: Cambridge University Press.

Schwartz, Peter, and Doug Randall. 2003. An Abrupt Climate Change Scenario and Its Implications for National Security, October. Pentagon report, U.S. Department of Defense, Washington, DC. <http://www.gbn.com/articles/pdfs/Abrupt%20Climate%20Change%20February%202004.pdf> (accessed August 16, 2010).

Scott, Daniel, and Geoff McBoyle. 2007. Climate Change Adaptation in the Ski Industry. *Mitigation and Adaptation Strategies for Global Change* 12(8):1411–1431. <http://www.springerlink.com/content/1381-2386/12/8/> (accessed August 13, 2010).

Scott, Daniel, Geoff McBoyle, and Brian Mills. 2003. Climate Change and the Skiing Industry in Southern Ontario (Canada): Exploring the Importance of Snowmaking as a Technical Adaptation. *Climate Research* 23:171–181.

Shulman, Seth. 2006. *Undermining Science: Suppression and Distortion in the Bush Administration*. Berkeley and Los Angeles: University of California Press.

Skirbekk, Gunnar. 1981. Nasjon og natur, eit essay om den Norske veremåten. In *Ord og bilde: En essaysamling*, ed. Asbjorn Aarnes. Oslo: Sstenersen, 13–22.

Skogen, Ketil. 1993. Risiko i et trygghetssamfunn: en hverdagshistorie fra Trysil. *Sosiologisk Tidsskrift* 9 (3):221–223.

Slovic, Paul, ed. 2000. *The Perception of Risk*. London: Earthscan.

Smith, Dorothy E. 1987. *The Everyday World as Problematic*. Boston: Northeastern University Press.

Statistisk sentralbyrå. 2002. Statistical Yearbook of Norway 2001. <http://www.ssb.no/aarbok/2001/tab/t-010410-034.html> (accessed August 14, 2010).

Sterman, John, and Linda Sweeney. 2007. Understanding Public Complacency about Climate Change: Adults Mental Models of Climate Change Violate Conservation of Matter. *Climatic Change* 80:213–238.

Stern, Paul, Thomas Dietz, and Gregory Guagnano. 1995. The New Ecological Paradigm in Social-Psychological Context. *Environment and Behavior* 27 (6):723–743.

Stoll-Kleemann, Susanne, Tim O'Riordan, and Carlo Jaeger. 2001. The Psychology of Denial Concerning Climate Mitigation Measures: Evidence from Swiss Focus Groups. *Global Environmental Change* 11:107–117.

Stotland, E., K. Mathews, S. Sherman, R. Hansson, and B. Richardson. 1978. *Empathy, Fantasy, and Helping*. Beverly Hills, CA: Sage.

Sturgeon, Noël. 1997. *Ecofeminist Natures: Race, Gender, Feminist Theory, and Political Action*. New York: Routledge.

Swidler, Anne. 1986. Culture in Action. *American Sociological Review* 51:273–286.

Swim, Janet, Susan Clayton, Thomas Doherty, Robert Gifford, George Howard, Joseph Reser, Paul Stern, and Elke Weber. 2009. *Psychology and Global Climate Change: Addressing a Multi-faceted Phenomenon and Set of Challenges*. Washington, DC: American Psychological Association.

Sydnes, A. K. 1996. Norwegian Climate Policy: Environmental Idealism and Economic Realism. In *Politics of Climate Change: A European Perspective*, ed. T O'Riordan and J Jäger. London: Routledge, 268–97.

Szaz, Andy. 2007. *Shopping Our Way to Safety: How We Changed from Protecting the Environment to Protecting Ourselves*. Minneapolis: University of Minnesota Press.

Thoits, Peggy. 1996. Managing the Emotions of Others. *Symbolic Interaction* 19:85–109.

de Tocqueville, Alexis. 1969. *Democracy in America*. Ed. J. P. Mayer. Trans. George Lawrence. New York: Anchor.

Ungar, Sheldon. 1992. The Rise and (Relative) Decline of Global Warming as a Social Problem. *Sociological Quarterly* 33 (4):483–501.

Ungar, Sheldon. 2003. Misplaced Metaphor: A Critical Analysis of the "Knowledge Society." *Canadian Review of Sociology and Anthropology/Revue Canadienne de Sociologie et d'Anthropologie* 40 (3):331–347.

Ungar, Sheldon. 2007. Public Scares: Changing the Issue Culture. In *Creating a Climate for Change: Communicating Climate Change and Facilitating Social Change*, ed. Susanne C. Moser and Lisa Dilling. New York: Cambridge University Press.

Ungar, Sheldon. 2008. Ignorance as an Under-identified Social Problem. *British Journal of Sociology* 59 (2):301–326.

United States Energy Information Administration (US EIA). 2009. Norway Energy Profile. <http://tonto.eia.doe.gov/country/country_energy_data.cfm?fips=NO&Go=Go> (accessed July 18, 2010).

U.S. Government Accountability Office (US GAO). 2004. Alaska Native Villages Affected by Flooding and Erosion Have Difficulty Qualifying for Federal Assistance (GAO-04-895T). <http://www.gao.gov/new.items/d04895t.pdf> (accessed August 13, 2010).

U.S. Government Accountability Office. 2007. Climate Change: Agencies Should Develop Guidance for Addressing the Effects on Federal Land and Water Resources (GAO-07-863). <http://www.gao.gov/products/GAO-07-863> (accessed August 12, 2010).

Vormeland, Oddvar. 1993. Education in Norway. In *Continuity and Change: Aspects of Contemporary Norway*, ed. Anne Cohel Kiel. Oslo: Scandinavia University Press, 205–220.

Watson, Robert T., Marufu C. Zinowera, and Richard H. Moss, eds. 1998. *The Regional Impacts of Climate Change: An Assessment of Vulnerability*. Cambridge, UK: Cambridge University Press.

Watt-Cloutier, Shelia. 2007. Testimony before the Inter-American Commission on Human Rights, Washington D.C. March 1. <http://www.earthjustice.org/news/press/2007/nobel-prize-nominee-testifies-about-global-warming> (accessed August 16, 2010).

Weart, Spencer. 2009. The Discovery of Global Warming. <http://www.aip.org/history/climate> (accessed August 12, 2010).

Weart, Spencer. Forthcoming. From Myth to Mandate in Problem Perception. In *Oxford Handbook of Climate Change and Society*, ed. John Dryzek, David Schlosberg, and Richard Norgaard. Oxford, UK: Oxford University Press.

Weiskel, Timothy. 2005. From Sidekick to Sideshow—Celebrity, Entertainment, and the Politics of Distraction: Why Americans Are "Sleepwalking toward the End of the Earth." *American Behavioral Scientist* 49 (3):393–409.

Wendler, Gerd, and Martha Shulski. 2009. A Century of Climate Change for Fairbanks, Alaska. *Arctic* 62 (3):295–300.

West, Cornel. 2000. "America's Night Side," speech given at the International Press Institute World Congress, Boston, May 2. <http://www.freemedia.at/fileadmin/media/Documents/Boston_2000_Congress_Report_01.pdf> (accessed September 7, 2010).

Williams, Simone J., and Gillian Bendelow. 1997. Introduction: Emotions in Social Life: Mapping the Sociological Terrain. In *Emotions in Social Life: Critical Themes and Contemporary Issues*, ed. Gillian Bendelow and Simon J. Williams. London: Routledge, xii–xxvii.

World Bank. 2010. *World Development Report 2010 Development and Climate Change*. Washington DC: The International Bank for Reconstruction and Development/The World Bank.

Zahran, S., S. Brody, H. Grover, and A. Vedlitz. 2006. Climate Change Vulnerability and Policy Support. *Society & Natural Resources* 19:1–19.

Zerubavel, Eviatar. 1997. *Social Mindscapes: An Invitation to Cognitive Sociology*. Cambridge, MA: Harvard University Press.

Zerubavel, Eviatar. 2002. The Elephant in the Room: Notes on the Social Organization of Denial. In *Culture in Mind: Toward a Sociology of Culture and Cognition*, ed. Karen Cerulo. New York: Routledge, 21–27.

Zerubavel, Eviatar. 2006. *The Elephant in the Room: Silence and Denial in Everyday Life*. New York: Oxford University Press.

Index

Aall, Carlo, 228
Access, gaining, 238–239
ACNielsen, 76
Advertising, 151–153
Affect, 65, 90, 214
Afghanistan War, 203
Agriculture, 15, 16, 113, 234, 240
   climate change impacts, 185, 188
   community-supported, 227
   mountains, 155
   skiing and, 17–18, 37
Air travel, 75, 221
Alaska, xv, 185, 187
Allemannsferdselrett (freedom of travel and land use), 147, 156
American Geophysical Union, 190, 191–192, 193
"Amerika as tension point" narrative, 142, 163–167, 171, 174–175
Amnesty International, 235
Andersen, Benedict, 116
Anti-intellectualism, 203
Apathy, xvi–xviii, 58–60, 60–62, 85. *See also* Guilt; Powerlessness
   American vs. Norwegian, 178–179
   efficacy and, 76, 101, 191–192
   guilt and, 195
   lack of political talk, 105, 215–216
   local news and, 51
   political alienation and, 79
   political economy and, 225
   powerlessness and, 190
   research explanations for, 1–3
   significance of, 13, 177–178
   social construction of, 33–34, 207
   study of, 63, 209
   as suffering, 222
Arendt, Hannah
   importance of conversation, 97–98, 122
   political talk, 41–43, 53
   volunteer organizing, 44
Attention. *See also* Attention, selective; Attention norms
   shifting of, 128–132
   social construction of, 6, 11
Attention, selective. *See also* Local focus; Past, connection to
   cultural narratives and, 173
   cultural tool kit, 123
   as emotion management, 174, 182, 197
   examples of, 132
   narratives, 213–214
Attention norms, 76, 112–120, 173
   as cultural tool, 121–122
   invisibility of, 132, 133
   local focus and, 25, 119–120
   organization of space, 116–117
   political economy, 210
   in production of denial, 212–213
   sense of past, 112–116, 118–119
Australia, 178
Avoiding Politics: How Americans Produce Apathy in Everyday Life (Eliasoph), 33, 215

Awareness
   American vs. Norwegian, 179
   vs. apathy, xvi–xviii
   vs. concern, 2, 68, 74, 177,
      191–192
   denial of self-involvement and,
      71–72
   emotion management and, 208
   emotions and, 223
   national identity and, 87–88
   powerlessness and, 195–197
   social organization of, 13, 72

Beck, Ulrich
   individualism, 204, 226
   risk societies, 33, 83, 84
   wealth inequality, 62
Belgium, 22
Bell, Michael, 150
Bellah, Robert, 192, 193, 225
Bergens Tidende (*Bergen Times*), 40, 42, 92
Berggreen, Brit, 149
Bondevik, Kjell Magne, 39
Bostrom, Ann, 64, 65
Brazil, 77
Brende, Børge, 71
Brewer, Thomas, 64
Brown, Molly Young, 58–59, 82
Brulle, Robert, 225
Brundtland, Gro Harlem, 71
Bryn, Steinar, 163, 165
Buell, Frederick, 177, 189
Bulkeley, Harriet, 63, 65, 72
Burke, Peter, 90–91
Bush, George W. (and administration of)
   Kyoto Protocol, 40, 45–46, 166,
      186, 202–203
   skepticism, xviii, 68–69, 186,
      233
Bygdaby, 245n1 (prologue)
   political engagement in, 16–17
   town description, 14–15, 26,
      234–235
Bygdaby Posten (newspaper). *See*
   Newspapers, local

Canada, 77
Capitalism, 62, 220–222
Carbon emissions. *See* Emissions
Carbon footprints, 77, 170, 228,
   247n3 (chap. 5)
Carbon trading, 140–141, 167, 172
Caring. *See* Concern
Chernobyl accident, 22–23
Church's Emergency Aid (Kirkens
   Nødhelp), 157
Citizenship and globalization, 222
City council, 99
Claims to virtue, 143, 162–163,
   171–175
Climate change
   assessment of, 40
   basic facts, 2
   climate models, 64
   as national issue, 43–44
   rising sea levels, xv, 76, 185, 221
Climate talks at The Hague, xvi–xvii,
   39, 79
Climate vs. weather, xvi
Coal, 2, 67, 172, 173
Coastlines, Florida, 185
Cognition, 214
   cognitive architecture, 210
   emotions and, 89–91, 91–92,
      93–94, 207
   political economy and, 12
   social, 11, 90, 207, 212
Cognitive dissonance. *See also*
   Double realities
   emotions and, 190, 194, 214
   identity and, 217
   individualism and, 204–205
   privilege and, 216
   self-efficacy and, 67–68, 90, 190,
      194, 217
Cognitive traditions, 5–6, 13, 14
Cohen, Stanley
   apathy, 33, 59
   denial and identity, 88
   doubleness, 60, 72
   interpretive denial, 162–163, 172
   invisibility of denial, 133
   literal denial, 179

modern capitalism, 220, 221
types of denial, 10–11, 211–212
Cold War, 164
Common Knowledge (Bulkeley), 63
Concern, 1, 73–74
   American lack of, 177, 184
   vs. awareness, 2, 68, 74, 177, 191–192
   decline in, 74, 78
   vs. emissions, 77
   vs. self-interest, 193
   social norms, 75
   United States, 74, 77, 187
   vs. wealth, 77
Confirmation bias, 65
Conforming
   equality and, 154
   guilt and, 75, 86–87, 195, 211
   masculinity and, 107
   social pressure, 75, 97, 122–123
Constitution Day, 24, 25, 81, 156
Consumerism. *See* Materialism
Control vs. powerlessness, 101
Conversation, 11, 52–62. *See also* Conversational control/tactics; Conversation norms; Small talk
   apathy, 59–60
   climate change as killer of, 55–57
   democracy and, 97–98, 122
   emotion management, 208
   meaning-making, 122
   political talk, 41–43, 53, 58, 99–101, 105, 201, 215–216
   small talk, 54
   sociological imagination, 11, 98, 121
Conversational control/tactics, 182. *See also* Conversation norms
   attention shifting, 128–132
   humor, 123–126
   information control, 126–127
   interpretive denial and, 104–105
   as tool of order, 171
   topic changing, 198
Conversation norms, 97–105, 210, 213. *See also* Conversational control/tactics
   apathy, 215–216

being intelligent, 166
educational talk, 101–102, 200
interviews, 98, 104–105
invisibility of, 132
political talk, 99–101
production of public sphere, 215–216
small talk, 98–99
social gatherings, 102–103, 200–201
Crystal Night, 17, 45, 245n1 (chap. 2)
Cultural homogeneity, 25–27, 31, 86, 160–161, 233
Cultural narratives. *See also* "Amerika as tension point"; "Norway is a little land"; "We have suffered"
   emotion management and, 213
   government and, 135, 140–141, 211
   interpretive denial and, 141–143, 202
   national identity and, 140–141
   nationwide applicability, 237
   privilege and, 218
   selective attention and, 173
   as tool of socially organized denial, 207, 215
   in the United States, 182, 202–203
Cultural tool kit, 7, 11, 215
   emotion management, 121–122, 213–214
   innocence, 222
   national identity and, 146
Cultural transformation, 247n1 (appendix)

Decision-making, 180
   affect and, 90, 214
   cognitive traditions and, 6
   faulty, 9, 65, 80
   individualism and, 226
   Iroquois Nation, 114
Deep acting, 93
Democracy, 74. *See also* Political alienation
   American national identity, 203
   vs. denial, 181–182, 207–208, 226

Democracy (cont.)
  equality and, 154
  globalization and, 204, 218, 222, 225–226
  importance of conversation, 97–98, 122
  local focus, 228
  Norwegian political system, 15
  political talk and, 43
  power and, 133
  volunteer organizing and, 44
Denial (individual), 6, 246n3 (chap. 2)
Denial (socially organized), 211–213, 214–215. *See also* Attention, selective; Attention norms; Cognitive dissonance; Conversation; Conversation norms; Cultural narratives; Double realities; Emotion management; Emotion norms; Emotions; Perspectival selectivity
  apathy and, 59, 60–62, 222
  breaking through, 223–224
  defined, 9–11, 13, 121
  vs. democracy, 181–182, 207–208, 226
  importance of understanding, 221
  vs. information deficit model, 12
  strength of model, 134
Denial, implicatory, 10–11, 181–182, 202, 211
Denial, interpretive, 10, 211
  claims to virtue and, 162–163, 172
  cultural narratives and, 141–143, 202
  examples of, 104–105
  guilt and, 133
Denial, literal, 10, 179, 181–182, 202, 211. *See also* Skepticism
Denial of outcome severity, 100, 105, 126, 246n2 (chap. 2)
Denial of self-involvement, 44, 71–72, 123, 142, 167
Denmark, 147–148
Den Norske væremåten (The Norwegian Way of Being) (Klausen), 158–159

Dialects, 247n3 (appendix)
  insider/outsider distinction, 24–25
  local focus and, 17, 117–118, 227
  site selection and, 237
  tradition and, 115, 153–154
Disease, xv, 22, 85, 221
Double realities, 3–7, 120–123. *See also* Cognitive dissonance
  commonness of, 204
  as denial, 60–62
  emotions and, 8, 11
  local focus and, 227, 228
  political economy and, 225
  privilege and, 218, 221–222
  in the United States, 186–187, 201
Drinking behavior, 102–103, 107, 239
Dugnad (volunteering), 23–24, 44–45, 155–156
Dunlap, Riley, 77
Durkheim, Émile, 210

Earth Summit (Rio de Janeiro, Brazil), 183
Ecologist, The, 178
Economic impacts
  ice fishing and, 20, 34–35
  lack of conversation, 57
  skiing and, xiii, xiv, xvii, 18–20, 34–35, 35–36
  in the United States, 184–186
Educators/educational talk, 98, 101–102, 105, 200. *See also* Schools
Efficacy/self-efficacy, 222. *See also* Powerlessness
  apathy and, 76, 101, 191–192
  as attention shifting, 123
  cognitive dissonance and, 67–68, 90, 190, 194, 217
  national identity and, 158
  need for hope, 224
  optimism and, 83, 127, 174, 198–199
  vs. powerlessness, 83–84, 91, 101, 128–129, 192–193, 194, 197, 198
  responsibility displacement and, 44

Egalitarianism
  national identity and, 28, 29–30, 88, 138, 141, 142, 145, 154–158
  vs. Norwegian wealth, 86, 88, 110, 143, 157, 162
  welfare state and, 31
Electric supplies, 185, 200
Eliasoph, Nina, 33, 228
  closeness to home, 76, 101
  conversation, 53, 59, 105, 215
  hegemony, 133–134
  humor, 125
  knowing and not knowing, 104
  local news, 51
Emissions, 10, 45, 46, 162, 211
  as climate change cause, 66, 73
  vs. concern, 77
  fires, 125
  government and, 39
  Norwegian wealth and, 12, 70–71
  per capita, 142, 170
  reduction targets, 167, 172–173, 224–225, 228, 246n1 (chap. 3)
  United States, 52, 142, 178, 203
Emotion management, 90–91, 91–95. *See also* Attention, selective; Cultural narratives; Efficacy/self-efficacy; Emotion norms; Information control; Perspectival selectivity
  apathy, 215–216
  awareness and, 208
  cultural tool kit, 121–122, 213, 213–214
  humor, 50, 123–126
  information control, 127, 134, 197–198
  optimism, 198–199
  perspectival selectivity, 163, 166
  political economy and, 210–211
  social norms, 123
  sociological imagination, 208
  spin, 162–163
  strategies of, 132–133, 174
  United States, 182
Emotion norms, 92–93, 106–112, 174. *See also* Identity, national; Maintaining control; Optimism
  being cool, 106, 109–110, 124
  being smart, 106, 110–112, 126
  in construction of everyday reality, 132, 210, 212–213, 215
  translation, 238
  in the United States, 200
Emotions, 7–9, 80, 174, 223. *See also* Affect; Fear; Guilt; Identity; Ontological security; Powerlessness
  ambivalence, 72
  cognition and, 89–91, 91–92, 93–94, 207
  cognitive dissonance and, 190, 194, 214
  democracy and, 207
  expressed vs. real, 247n2 (appendix)
  implicatory denial and, 181
  political economy and, 10
  power reproduction and, 97, 208, 214
  significance of, 134, 210, 213
  social movements and, 8–9, 11, 94, 95
  United States, 187–197
Emotions, sociology of, 11, 90, 92, 210, 213–215
Empathy, 59, 61
Empowerment, 224
England, xvii, 120, 150, 178
Environmental justice, 194, 213
  national identity and, 88
  privilege and, 86, 216, 219, 221–222
  uneven climate change effects, 70
Environmental Protection Agency, 183
Environmental sociology, 209–210, 213, 219
Environment Poll (Gallup), 184, 187
Equality. *See* Egalitarianism

Eriksen, Thomas Hylland
  connection to nature, 137–138, 146
  contested national identity, 141
  egalitarianism, 29
  national identity elements, 138, 142, 146
  simplicity, 158
Essentialism, Norwegian, 139, 145
Europe, xv, 186. *See also specific countries*
European Union, opposition to, 17, 18, 80, 117
Evolution, 203
Exceptionalism, American, 202–203, 218
Exceptionalism, Norwegian, 165
Extinction, 111, 221

Faarlund, Nils, 150
Farmers/farms. *See* Agriculture
Farm Women's Association, 132, 235
Fear, 8, 177, 188–189. *See also* Ontological security; Powerlessness
  vs. awareness, 87, 89, 223
  vs. connection with nature, 148
  display rules for, 92
  dissonance and, 214
  management of, 174
  masculinity and, 200
  ontological security, 82
  powerlessness and, 197–198
  private expression of, 103
Festinger, Leon, 67–68
Finland, 139
Fishing. *See* Ice fishing
Flooding
  Alaska, xv
  England, xvii, 39, 120
  environmental justice and, 216
  Norway, xiii, xvi, 37, 39
Florida, 185
Food insecurity, xv, 221
Foods, traditional, 24, 115, 160
Forbes.com, 179
Formality, 237, 239

Framework Convention on Climate Change, 183
France, 247n3 (chap. 5)
Frantz, Cynthia, 68, 246n3 (chap. 2)
Freedom of travel and land use (*allemannsferdselrett*), 147, 156
Fremmed skeptis (skepticism of strangers), 237, 239
Freudenburg, William, 219
Friluftslivet ("open-air life"), 147, 150
Future. *See also* Past, connection to; Time, sense of
  avoidance of, 83, 115, 128–129
  sense of, 76, 80

Gamson, William, 46
Gecas, Victor, 90–91
Generational differences, 247n1 (appendix)
Germany, 247n3 (chap. 5)
Giddens, Anthony
  globalization, 62, 219
  individualism, 204, 226
  lack of climate change politics, 191, 224
  ontological security, 26, 81–82
Globalization, 62, 227
  democracy and, 204, 218, 222, 225–226
  privilege and, 62, 220–222
  time-space distanciation, 219
Global warming. *See* Climate change
Goffman, Erving, 102, 142, 162
Gordon, Hava, 124, 215
Gore, Al, xviii, 180, 188
Government, 78. *See also* Norway; Welfare state
  cultural homogeneity and, 27, 160–161
  emission reduction targets, 228
  failure of planning, 200
  inadequacy of, 84, 190–191, 192–193
  Kyoto Protocol, 70, 71
  narratives, 135, 140–141, 211
Government Accountability Office (U.S.), xv, 184

Gramsci, Antonio, 7, 11, 133–134, 218
Greenhouse gas emissions. *See* Emissions
Greeting norms, 108
Grendelag (neighborhood associations), 23–24
Groven, Kyrre, 228
Guilt, 80, 85–87, 194–197
  vs. awareness, 223
  conforming and, 75, 86–87, 195, 211
  connection with nature and, 151
  dissonance and, 90, 194
  humanitarianism and, 157
  interpretive denial and, 133
  management of, 174–175, 199
  national identity and, 86, 194
  political economic conditions, 10
  privilege and, 86, 195, 217
  selective attention, 197
  shifting blame, 166
Gulf Stream, xv
Gullestad, Marianne, 107–108, 160

Habermas, Jürgen, 41–43, 44, 97–98
Hagen, Kåre, 171
Hague, The. *See* Climate talks at The Hague
Halford, Grame, 1, 65
Hansen, James, 183, 184
Harald V (King of Norway), 39–40, 99
Heat islands, urban, xv
Heat waves, xv
Hegemony, 133
Hellevik, Ottar, 2, 78
Helping behavior, model of, 68, 246n3 (chap. 2)
Hiking/hiking clubs, 146–147, 148, 231. *See also* Skiing
Hippe, Jon, 171
Hiroshima survivors, 4
History. *See* Memory; Past, connection to; World War II
Hobsbawm, Eric, 116, 129, 130

Hochschild, Arlie, 8, 92, 93–94, 121–122
Hollander, Jocelyn, 124, 215
Home life, 103
Hovden, Eivind, 10, 71, 172, 173
Humanitarianism, 141, 143, 145, 154–158
  aid levels, 170, 215
  as Norwegian value, xviii, 31, 87–88
  privilege, 155
  religious beliefs and, 28
Humor, 50, 51, 123–126, 199

Ice, lake, xiv, 34–35
Ice, sea, 185
Ice fishing, xiii, xiv, 20, 34–35
Identity. *See also* Cognitive dissonance; Ontological security
  goodness, 150–151, 158
  importance of, 80, 90, 207, 211, 217
  tradition and, 130–131
Identity, national, 27, 87–89. *See also* Cultural narratives; Simplicity
  connection to nature, 20–21, 29, 137–140, 141, 142, 146–151, 162
  egalitarianism, 28, 29–30, 88, 138, 141, 142, 145, 154–158
  elements of, 13–14, 138, 142, 143–145, 161–162
  guilt and, 86, 194
  narratives of, 140–141
  pride in, 174–175
  rural life, 138, 141, 142, 151–154, 162
  stereotypes, 140, 246n1 (chap. 5)
  as tool of order, 145–146, 171, 173–175
Ignoring, 97, 121, 197, 213–214
Imagined community, 116, 173
Immerwahr, John, 55
Impression management, 162, 172
Inaction. *See* Apathy
Inconvenient Truth, An (Gore), xviii, 180, 188

Independence, 106–107
Independent, The, 177–178
Indigenous people, xv, 185, 187
Individualism, 191, 192–193, 204–205, 226
Industry, power of, 193–194
Information control, 126–127, 134, 174, 196, 197–198
Information deficit model, 12, 63, 64–67, 212
  inadequacy of, 1–2, 3, 67, 69–70, 72
Innocence. *See also* Responsibility, diffusion/deflection of; Tools of innocence
  construction of, 133, 213, 222
  cultural narratives and, 215
  denial of self-involvement, 71–72
  of nature, 162
  perspectival selectivity and, 165, 170
  privilege and, 218
  simplicity and, 161
Insider/outsider distinction, 31
  connection to nature and, 148
  connection to past and, 114
  dialects and, 24–25
  formality and, 237
  local focus and, 117
  national identity and, 145–146
Insurance, 38, 185–186
Inter-American Commission on Human Rights, 185
Intergovernmental Panel on Climate Change (IPCC), 183
Interviews
  conversation norms in, 98, 104–105
  list of people, 243–244
  methods, 234, 239–241, 241–242, 245n2 (prologue)
Inuit people, xv, 185
Invention of Tradition, The (Hobsbawm and Ranger), 129
Iraq War, 203
Iroquois Nation, 114
Isolationism, 117

Jacques, Peter, 202
Jaeger, Carlo, 81, 84, 214
James, Henry, 220
Jante's Law, 154
Jasper, James, 8–9, 31
Jonassen, Christen, 20, 31

Katrina, Hurricane, xvii, xviii, 179–180, 189
Kellstedt, Paul, 2, 68
Kent, Jennifer, 192–193, 204, 226
Kiel, Anne, 236
Kimmel, Michael, 218
Kirkens Nødhelp (Church's Emergency Aid), 157
Knowledge. *See* Awareness
Krosnick, Jon, 2, 68, 190
Kyoto Protocol, 88, 172, 183
  Bush and, 40, 45–46, 166, 186, 202–203
  JUSSCANNZ group, 245n2 (introduction)
  Norwegian government and, 70, 71

Labor Party (Norway)
  humanitarian work, 157
  oil expansion, 39, 46, 172
  welfare state and, 15–16
Labor Party, local, 43, 58, 99, 100, 158, 235
Lafferty, William, 71
Lakoff, George, 225
Langer, Ellen, 91
Larson, Peter, 185
Lean, Geoffrey, 178
Leiserowitz, Anthony, 180, 187, 214
Lertzman, Renee, 178
Lievanos, Raoul, 219
Lifton, Robert J., 222
  awareness, 223
  claims to virtue, 143, 163, 171
  double reality, 4–5, 11
  nature, 82
  psychic numbing, 4–5, 11, 89, 207, 216, 223
Lindseth, Gard, 10, 71, 172, 173, 228

Local focus
  apathy and, 58
  community building, 23–24
  democracy and, 228
  local newspapers and, 51–52, 119–120
  local patriotism, 17, 117–118, 227
  maintenance of, 25, 100–101, 129–132, 200
  place names, 113, 118–119
  vs. powerlessness, 87
  social movements, 227–229
Lorenzoni, Irene, 65, 214
Love, 107–108
Lukes, Steven, 133
Lutheran Church, 27, 159–160

Machado, Antonio, 227
Macy, Joanna, 58–59, 82
Mad cow disease, 85
Maintaining control, 106–109, 125, 211
  encouragement to, 110
  information control and, 126, 174
  local newspaper appeal to, 51
  masculinity and, 111, 200
  national identity and, 161
Managed Heart, The: Commercialization of Human Feeling (Hochschild), 8
Maniates, Michael, 226
Marx, Karl, 210
Masculinity, 106–107, 110–111, 200
Materialism
  antimaterialism, 27–28, 29, 160
  vs. citizenship, 192
  increasing, 161, 166
Mayer, Stephan, 68, 246n3 (chap. 2)
McCarthy, Michael, 177–178
McKibben, Bill, 189
Media. *See also* Newspapers, local; Newspapers, national; Television
  cultural homogeneity and, 233
  information deficit model and, 1, 64, 65, 66, 212
  narrative propagation, 135, 140–141

Meijnders, Anneloes, 214
Memory. *See also* Past, connection to; World War II
  monuments and, 114
  socially organized, 11, 116, 119, 212
  vulnerability and, 23
Men, 127
Methods, 231–242
  American counterpoint, 241–242
  choice of for research, 232–234
  gaining access, 238–239
  interviews, 234, 239–241, 241–242, 245n2 (prologue)
  participant observation, 234–236
  site selection, 236–238
Mexico, 77
Midden, Cees, 214
Military, United States, 186
Mills, C.W., 43, 76, 222
Mitigation vs. prevention, 228
Mjøset, Lars, 171
Moral deterioration, 193, 198
Moral imagination, 213, 222
Moral order. *See* Identity, national; Traditions
Moral order, winter weather and, 37
Moser, Susanne, 224, 228–229
Mountains
  cultural significance of, 20, 28, 147–148
  egalitarianism and, 155
  simplicity and, 150–151, 159
Myers, Teresa, 64
"Mythic Norway." *See* Identity, national

National Academy of Sciences Committee on Abrupt Climate Change, 183
National boundaries, 116
National Center on Atmospheric Research, 183
National costumes
  community building and, 24, 25
  connection to past and, 115, 130, 137, 138, 153

National identity. *See* Identity, national
National security, United States, 186
Nation-states, 222–223
Nature, connection to
 national identity and, 20–21, 29, 137–140, 141, 142, 146–151, 162
 as a refuge, 102
 romanticization of, 149–150, 162
 as tool of order, 148, 171
Needs, hierarchy of, 74–77
Neighborhood associations (*grendelag*), 23–24
Netherlands, 76
New Hampshire, xiv
Newport, Frank, 177, 180
Newspapers, local, 46–52, 117, 119–120, 235, 236
Newspapers, national, xvi, 38–41, 42, 236
Newspapers, United States, 179, 183
New York Times, 183, 184
Nicholsen, Shierry, 59, 197, 217
Nilsen, Ann, 76, 87, 115, 128
Nisbet, Matthew, 64
Nisselue (ski hat), 117, 158–159
Nobel Peace Prize, 155–156, 170
Nonparticipation. *See* Apathy
Norway. *See also* Emissions; Government; Identity, national; Oil industry, Norwegian; Wealth, Norwegian; Welfare state
 border closing, 144
 carbon footprint, 247n3 (chap. 5)
 choice of for research, 232–234
 environmental degradation, 144–145
 political participation, 16–17, 43, 79–80, 233
 State Petroleum Fund, 10
"Norway is a little land," 142, 160, 163, 169–171, 174–175
Norwegian Mountain Bread (Norsk Fjellbrod), 151, 152
Norwegian sensibility. *See* Identity, national

Norwegian Sustainable Development profile, 71
Norwegian Way of Being (*Den Norske væremåten*) (Klausen), 158–159
Nuclear threat, 186
Nutter, Franklin W., 185–186

O'Connor, Robert, 77
Oil industry, Norwegian
 claims to virtue, 172–173
 expansion of, 39, 46, 88, 140–141, 167, 171–172
 exploitation of third world, 144
 Norwegian wealth and, 9–10, 70–71, 142, 159, 167–168, 211
Oil industry, United States, xviii, 66, 68–69, 202
Ontological security
 cultural homogeneity and, 26
 maintenance of, 119, 120, 148
 threats to, 80, 81–83, 125, 197
 tools of order and, 146, 215
 winter weather and, 37
"Open-air life" (*friluftsliv*), 147, 165
Operasjon Dagsverk (Operation Day's Work), 157
Opotow, Susan
 denial of outcome severity, 100, 105, 126, 246n2 (chap. 2)
 denial of self-involvement, 44, 71–72, 123, 142, 167
Optical socialization, 5, 7, 112, 212
Optimism, 108
 declining, 78–79, 84
 in educational talk, 101–102, 105, 129
 efficacy and, 83, 127, 174, 198–199
 faith in government, 78
 information control and, 127, 174, 197–198
O'Riordan, Tim, 214
Oslo, 234
Oslo Accords, 155, 170

Outsider status of researcher, 234, 236, 238, 239
Ozone hole, 1, 65, 66, 84–85

Participant observation, 234–236
Past, connection to, 130–131, 154, 171
    attention norms and, 112–116, 118–119
    national costumes, 115, 130, 137, 138, 153
    skiing and, 113, 137, 138–140
Patriotism, local, 17, 117–118, 227
Pentagon, 186
Permafrost, 185
Perspectival selectivity, 94, 163–171, 174–175
    "Amerika as a tension point," 163–167
    narratives and, 141, 142, 163, 213–214
    national identity and, 173–174
    "Norway is a little land," 169–171
    tools for, 123
    "We have suffered," 167–169
Pew Research Center, 180, 184
Philippines, 247n3 (chap. 5)
Place, sense of, 51, 171, 187, 199–200
Place names, 113, 118–119
Polar ice caps, 221
Polar regions, xiv–xv
Political alienation, 79–80
    in the United States, 17, 79, 179, 191, 204, 225–226
Political organizing, 41–45
Political participation, 16–17, 43, 79–80, 233
Political parties, 99
Political talk, 41–43, 58, 99–101, 105, 201, 215–216. *See also* Conversation
Portugal, 77
Power, 11, 43, 53, 121. *See also* Power, reproduction of; Powerlessness

Power, reproduction of, 6–7, 133–134, 187
    attention norms and, 212
    emotions and, 97, 208, 214
    privilege and, 218
Powerlessness, xix, 54, 83–85. *See also* Political alienation
    awareness and, 195–197
    conversational control and, 125
    vs. efficacy, 83–84, 91, 101, 128–129, 192–193, 194, 197, 198
    individualism and, 204–205
    information control and, 198
    local focus and, 87
    management of, 166, 174
    "Norway is a little land" narrative and, 169–171
    vs. optimism, 101–102
    political economy and, 10
    vs. political talk, 43
    psychic numbing and, 89
    as threat to identity, 80, 211
    in the United States, 189–194
Precipitation, changes in, 185
Prevention vs. mitigation, 228
Privilege, 213, 216–220
    benefits of climate denial and, 72
    environmental justice and, 86, 216, 219, 221–222
    globalization and, 62, 220–222
    guilt and, 86, 195, 217
    hierarchy of needs, 75, 77
    humanitarian work and, 155
    inequity and, 86, 208
    maintenance of, 216
    normalizing of, 168
    research choices and, 232–233
Program on International Policy Attitudes, 178–179
Progress Party (Norway), 144
Project Omelas, 178
Property, private, 147, 156
Psychic numbing, 4–5, 11, 89, 207, 216, 223
Pulido, Laura, 219

Racism, 45, 89, 144, 162, 174
Ranger, Terence, 116, 129, 130
Read, Daniel, 1–2, 67
Reinsurance Association of America, 185–186
Religion. *See* Lutheran Church
Renn, Ortwin, 81
Research methods. *See* Methods
Responsibility, diffusion/deflection of, 68. *See also* Innocence; Tools of innocence
  cognitive dissonance and, 214
  denial of self-involvement, 44, 123
  Norwegian exceptionalism and, 165
Risk models, 9
Risk perception, 65, 214
Risk societies, 33, 80, 83, 84, 202, 204
Rosa, Eugene, 81
Rosenberg, Morris, 90, 91, 141, 163, 173
Rotary Club, 235
Roy, Arundhati, 178
Rural life, 138, 141, 142, 151–154, 162
Rural-urban differences, 247n1 (appendix)
Russia, 21–23, 77, 116, 143, 157, 168

Sampling, maximum variation, 235
Sandvik, Hanno, 12, 77
Schneider, Stephen, 183
Schools, 103, 165, 233. *See also* Educators/educational talk
  standardized curriculum, 26, 156
Science communication, 9
Science of climate change, 179
  lack of trust, 202, 203
  manipulation of, 64
  United States leadership, 183–184
Sea levels, rising, xv, 76, 185, 221
Selective attention. *See* Attention, selective
Selective interpretation, 94, 141–142
Self-efficacy. *See* Efficacy/self-efficacy

Self-righteous comparisons, 167. *See also* "Amerika as tension point"
Sheehan, Peter, 1, 65
Simplicity, 138, 140, 141, 158–161
  connection to nature and, 150, 162
  overcommunication of, 142
Site selection, 236–238
Skepticism, 2, 186, 233. *See also* Denial, literal
  campaigns of, xviii, 66, 68–69, 181–182, 203
  decline in, 68–69
  humor and, 199
  inhibition of political talk, 201
  mistrust of climate science, 179, 202
Skiing, xiii, xiv, xvii
  in advertisements, 151–152
  agriculture and, 17–18, 37
  artificial snow, 51, 57
  connection to nature and, 146–147
  cultural significance of, 17–18, 19, 36–37, 76
  local media coverage of, 46, 47–48
  low snowfall and, 18–20, 35–36
  moral order, 148
  national identity and, 29, 158
  sense of past and, 113, 137, 138–140
  ski clubs, 52, 113, 231, 246n2 (chap. 5)
  United States, 185
Skirbekk, Gunnar, 20, 150
Skogen, Ketil, 5, 150
Slovic, Paul, 90
Slow Food Movement, 227
Small talk, 54, 58, 98–99, 105
  weather, xiii, xvi
Smith, Dorothy, 33, 41, 210
Snow, artificial, xiv, 35–36, 47–48, 51, 57
Snowfall, low, xiii, xv, 34, 52, 81
  economic impacts of, 18–20, 35–36
  snowpack, 185

Social gatherings, 98, 102–103
Social Mindscapes (Zerubavel), 13, 112
Social movements
  emotion management, 213
  emotions, 8–9, 11, 94, 95
  moral voice, 31
  motivation for, 196–197
  nonparticipation, 58–60
  return to the local, 227–229
Social networks, 83
Sociological imagination, 76
  conversation and, 11, 98, 121
  emotion management, 208
  emotions and, 8
  local focus and, 100–101, 228
  local newspapers and, 47, 51
  political talk and, 43, 53
Sociology of cognition, 112
Sondagstur (Sunday hike), 52, 231
SOS Barneby, 157
SOS Children's Village, 245n1 (chap. 1)
Spain, 247n3 (chap. 5)
State Petroleum Fund (Norway), 10
States of Denial: Knowing about Atrocities and Suffering (Cohen), 33, 211–212
Sterman, John, 1, 64
Stoll-Kleemann, Susanne, 214
Stoltenberg, Jens, 39, 40, 99, 167
Suffering narrative, 142, 171, 218
Surface acting, 93
Sweden, 17, 147–148, 247n3 (chap. 5)
Sweeney, Linda, 1, 64
Swidler, Ann, 7, 11, 121, 146, 215

Taxes, 160
Television
  cultural homogeneity and, 26, 233
  national identity and, 161
  research methods and, 235, 236
  in the United States, 180
  volunteering and, 156

Temporal containment of harm, 100, 246n2 (chap. 2)
Thought communities, 6, 112, 187
Time, sense of, 51, 114, 115. *See also* Future; Past, connection to
Time magazine, xiv–xv, 180
Time-space distanciation, 219–220, 226
Tipping point, 189
Tocqueville, Alexis de, 192, 225
Tools of innocence, 11–12, 171, 213, 226. *See also* Cultural narratives; Innocence; Responsibility, diffusion/deflection of
  connection to nature and, 148, 149–150
  distance from responsibility, 146, 215
  egalitarianism and humanitarianism, 157–158
Tools of order, 11–12, 171, 213, 226
  connection to nature, 148, 171
  national identity, 145–146, 171, 173–175
  ontological security and, 146, 215
  traditions, 129–132, 171
Toughness. *See* Maintaining control
Tourism, 15, 20, 131, 137. *See also* Ice fishing; Skiing
Toxic waste, 219–220
Traditions, 21. *See also* National costumes
  food, 24, 115, 160
  increased interest in, 154
  invented, 116, 129, 130, 232
  local focus and, 120
  moral order and, 215
  national identity and, 145–146
  preparations of sheep's heads, 14
  rural life and, 153
  seasonal ritual, 49
  sense of past and, 113, 115
  simplicity and, 160
  social networks and, 23–24
  storytelling as, 27

Traditions (cont.)
  as tools of order, 129–132, 171
  variation across Norway, 237
Translation, 238
TV Aksjon (TV Action), 156, 157–158
Typisk Norsk? (Typical Norwegian?), 246n1 (chap. 5)

Umbrella Group, 10, 88, 245n2 (introduction)
United Nations, 39, 40
United States. *See also* "Amerika as tension point"; Bush, George W.; Skepticism
  acknowledgment of climate change in, 179–180
  carbon footprint, 247n3 (chap. 5)
  climate impacts on, 184–185
  climate science leadership, 183–184
  concern in, 74, 77, 187
  as cultural model, 164
  cultural narratives in, 182, 202–203
  discomfort discussing climate change, 55
  emissions, 52, 142, 170, 178, 194, 203
  environmental degradation, 145
  exceptionalism, 177
  exploitation by, 144
  fear in, 188–189
  guilt in, 194–197
  Hurricane Katrina, xvii, xviii, 179–180, 189
  implicatory denial in, 181–182
  individualism in, 204–205
  Inuit people and, xv
  Kyoto Protocol, 40, 45–46, 166, 186, 202–203
  as negative example, 46, 108, 160
  Norwegian immigration, 164
  oil industry, xviii, 66, 68–69, 202
  political alienation in, 17, 79, 179, 191, 204, 225–226
  powerlessness in, 189–194
  research methods in, 241–242
  wilderness movement, 150
Urbanization, 153

Vedlitz, Arnold, 2, 68
Vermont, xiv
Virtue, claims to, 143, 162–163, 171–175
Volunteering (*dugnad*), 23–24, 44–45, 155–156
Vormeland, Oddvar, 156
Vulnerability, 21–23, 108, 128

Walser, Maggie, 224
Water, xv, 185, 188, 200
Watt-Cloutier, Shelia, 185
Wealth, Norwegian, 12, 30
  vs. egalitarianism, 86, 88, 110, 143, 157, 162
  environmental degradation, 145
  narrative of suffering, 168–169
  oil industry and, 9–10, 70–71, 142, 159, 167–168, 211
Wealth vs. concern, 77
Weart, Spenser, 180
Weather differences, xv–xvi, 66–67, 69, 81, 104
Weather events, extreme. *See* Flooding; Katrina, Hurricane
Weather vs. climate, xvi
Webler, Thomas, 81
"We have suffered" narrative, 142, 171, 218
Weiss, Leah
  denial of outcome severity, 100, 105, 126, 246n2 (chap. 2)
  denial of self-involvement, 44, 71–72, 123, 142, 167
Welfare state, 15–16, 155, 233
  decline of, 116
  egalitarianism and, 156
  justice and, 31
  pride in, 158
West, Cornel, 220
Wildfires, 185
Wilke, Henk, 214

World Health Organization, xv
World War II
  bombing of Bygdaby, 14, 23, 168
  monuments to, 114
  national identity and, 28
  occupation of Bygdaby, 15, 23, 119, 168
  suffering narrative and, 168–169
  United States and, 164
Worry. *See* Concern

Zahran, Sammy, 2, 68, 77
Zerubavel, Eviatar, 9, 134
  cognitive traditions, 13, 14
  concern, 75
  optical socialization, 5, 112, 212
  selective attention, 5–6, 173